Methods in Molecular Biology

Series Editor
John M. Walker
School of Life and Medical Sciences
University of Hertfordshire
Hatfield, Hertfordshire, UK

For further volumes:
http://www.springer.com/series/7651

For over 35 years, biological scientists have come to rely on the research protocols and methodologies in the critically acclaimed *Methods in Molecular Biology* series. The series was the first to introduce the step-by-step protocols approach that has become the standard in all biomedical protocol publishing. Each protocol is provided in readily-reproducible step-by step fashion, opening with an introductory overview, a list of the materials and reagents needed to complete the experiment, and followed by a detailed procedure that is supported with a helpful notes section offering tips and tricks of the trade as well as troubleshooting advice. These hallmark features were introduced by series editor Dr. John Walker and constitute the key ingredient in each and every volume of the *Methods in Molecular Biology* series. Tested and trusted, comprehensive and reliable, all protocols from the series are indexed in PubMed.

Chemogenomics

Methods and Protocols

Edited by

Daniel Merk

Department of Pharmacy, Ludwig-Maximilians-Universität München, Munich, Germany

Apirat Chaikuad

Institute of Pharmaceutical Chemistry, Goethe University Frankfurt, Frankfurt am Main, Germany

Humana Press

Editors
Daniel Merk
Department of Pharmacy
Ludwig-Maximilians-Universität
München
Munich, Germany

Apirat Chaikuad
Institute of Pharmaceutical Chemistry
Goethe University Frankfurt
Frankfurt am Main, Germany

ISSN 1064-3745 ISSN 1940-6029 (electronic)
Methods in Molecular Biology
ISBN 978-1-0716-3396-0 ISBN 978-1-0716-3397-7 (eBook)
https://doi.org/10.1007/978-1-0716-3397-7

This Humana imprint is published by the registered company Springer Science+Business Media, LLC, part of Springer Nature.
The registered company address is: 1 New York Plaza, New York, NY 10004, U.S.A.

Preface

In the search for new medicines, annotation of protein functions is a key step for the identification and validation of new targets mediating desired therapeutic effects. Chemogenomics, which studies the response of a biological system to sets of bioactive compounds, is a fruitful approach to this task in the postgenomic era. At an interface of chemistry and biology, it is a multidisciplinary field benefiting from a vast available number of highly annotated bioactive compounds and high-throughput phenotypic assays. This book brings together theoretical guidance and protocols on multiple aspects of the diverse discipline of chemogenomics covering considerations on chemogenomics library assembly, compound profiling and phenotypic assays. As an introduction to the wide field, the book starts with two chapters on the idea of chemogenomics (Chap. 1) and on the assembly and use of the Kinase Chemogenomics Set (KCGS) as a prime example (Chap. 2). Chapters 3, 4, 5, and 6 provide protocols on data mining for chemogenomics compound candidates (Chap. 3) and compound quality control (Chaps. 4, 5, and 6) as important general methods associated with chemogenomics. Chapters 7, 8, and 9 report protocols for protein family focused assay systems to profile chemogenomics compounds for selectivity which is a critical quality aspect for chemogenomics sets. Protocols in Chaps. 10, 11, 12, 13, 14, and 15 refer to diverse functional and target engagement assays in cellular settings for broad characterization of chemogenomics compounds, and Chaps. 16 and 17 describe methods for phenotypic assays in which chemogenomics sets may be applied. This diverse collection of protocols on the topics of chemogenomics has been made possible through the generous contributions of many experts in the field for which we are very grateful. We are also very grateful to the series editor John Walker and his guidance in preparing this book.

Munich, Germany — *Daniel Merk*
Frankfurt am Main, Germany — *Apirat Chaikuad*

Contents

Contributors

LAURIE AILLES • *Princess Margaret Cancer Centre, University Health Network, Toronto, ON, Canada; Department of Medical Biophysics, University of Toronto, Toronto, ON, Canada*

SAMIR H. BARGHOUT • *Structural Genomics Consortium, University of Toronto, Toronto, ON, Canada; Department of Pharmacology & Toxicology, Faculty of Medicine, University of Toronto, Toronto, ON, Canada*

DALIA BARSYTE-LOVEJOY • *Structural Genomics Consortium, University of Toronto, Toronto, ON, Canada; Department of Pharmacology & Toxicology, Faculty of Medicine, University of Toronto, Toronto, ON, Canada*

LORENE BRUNELLO • *Institute of Biochemistry 2, Faculty of Medicine, Goethe University Frankfurt, Frankfurt am Main, Germany; Buchmann Institute for Molecular Life Sciences, Goethe University Frankfurt, Frankfurt am Main, Germany*

SARA CANO-FRANCO • *Institute of Biochemistry 2, Faculty of Medicine, Goethe University Frankfurt, Frankfurt am Main, Germany; Buchmann Institute for Molecular Life Sciences, Goethe University Frankfurt, Frankfurt am Main, Germany*

APIRAT CHAIKUAD • *Institute of Pharmaceutical Chemistry, Goethe University Frankfurt, Frankfurt, Germany*

ALESSIO CIULLI • *Centre for Targeted Protein Degradation, Division of Biological Chemistry and Drug Discovery, School of Life Sciences, University of Dundee, James Black Centre, Dundee, UK*

NICHOLAS D. P. COSFORD • *NCI-Designated Cancer Center, Sanford Burnham Prebys Medical Discovery Institute, La Jolla, CA, USA*

ADAM P. CRIBBS • *Botnar Research Centre, NIHR BRC Oxford, Nuffield Department of Orthopaedics, Rheumatology and Musculoskeletal Sciences, University of Oxford, Oxford, UK; Oxford Translational Myeloma Centre, University of Oxford, Oxford, UK*

LAURENT J. S. DE BACKER • *NCI-Designated Cancer Center, Sanford Burnham Prebys Medical Discovery Institute, La Jolla, CA, USA*

ELIZABETH R. DOMINGUEZ • *Promega Corporation, Madison, WI, USA*

DAVID H. DREWRY • *Structural Genomics Consortium (SGC), UNC Eshelman School of Pharmacy, University of North Carolina at Chapel Hill (UNC-CH), Chapel Hill, NC, USA; Division of Chemical Biology and Medicinal Chemistry, UNC Eshelman School of Pharmacy, UNC-CH, Chapel Hill, NC, USA*

KAITLIN K. DUNN HOFFMAN • *Promega Corporation, Madison, WI, USA*

LEWIS ELSON • *Institute of Pharmaceutical Chemistry, Goethe University Frankfurt, Frankfurt, Germany; Structural Genomics Consortium, BMLS, Goethe University Frankfurt, Frankfurt, Germany*

MATTHIAS GSTAIGER • *Institute of Molecular Systems Biology at ETH Zurich, Zurich, Switzerland*

YE NA HAN • *NCI-Designated Cancer Center, Sanford Burnham Prebys Medical Discovery Institute, La Jolla, CA, USA*

HUNG HO-XUAN • *Institute of Biochemistry 2, Faculty of Medicine, Goethe University Frankfurt, Frankfurt am Main, Germany; Buchmann Institute for Molecular Life Sciences, Goethe University Frankfurt, Frankfurt am Main, Germany*

KILIAN V. M. HUBER • *Centre for Medicines Discovery, Nuffield Department of Medicine, University of Oxford, Oxford, UK; Target Discovery Institute, Nuffield Department of Medicine, University of Oxford, Oxford, UK*

LAURA ISIGKEIT • *Institute of Pharmaceutical Chemistry, Goethe University Frankfurt, Frankfurt, Germany*

KUNAL KARVE • *Princess Margaret Cancer Centre, University Health Network, Toronto, ON, Canada; Structural Genomics Consortium, University of Toronto, Toronto, ON, Canada*

STEFAN KNAPP • *Institute for Pharmaceutical Chemistry, Structural Genomics Consortium, Johann Wolfgang Goethe-University, Frankfurt am Main, Germany; Institute of Pharmaceutical Chemistry, Goethe University Frankfurt, Frankfurt, Germany; Structural Genomics Consortium, BMLS, Goethe University Frankfurt, Frankfurt, Germany*

NGAN LAM • *Promega Corporation, Madison, WI, USA*

LESTER J. LAMBERT • *NCI-Designated Cancer Center, Sanford Burnham Prebys Medical Discovery Institute, La Jolla, CA, USA*

DANIEL MERK • *Institute of Pharmaceutical Chemistry, Goethe University Frankfurt, Frankfurt, Germany; Department of Pharmacy, Ludwig-Maximilians-Universität München, Munich, Germany*

MONIQUE P. C. MULDER • *Department of Cell and Chemical Biology, Leiden University Medical Centre, Leiden, The Netherlands*

SUSANNE MÜLLER • *Institute of Pharmaceutical Chemistry, Goethe University Frankfurt, Frankfurt, Germany; Structural Genomics Consortium, BMLS, Goethe University Frankfurt, Frankfurt, Germany*

VÁCLAV NĚMEC • *Institute for Pharmaceutical Chemistry, Structural Genomics Consortium, Johann Wolfgang Goethe-University, Frankfurt am Main, Germany*

AMANDA N. NIEMAN • *Promega Corporation, Madison, WI, USA*

UDO OPPERMANN • *Botnar Research Centre, NIHR BRC Oxford, Nuffield Department of Orthopaedics, Rheumatology and Musculoskeletal Sciences, University of Oxford, Oxford, UK; Oxford Translational Myeloma Centre, University of Oxford, Oxford, UK*

DOMINIC D. G. OWENS • *Structural Genomics Consortium, University of Toronto, Toronto, ON, Canada*

STEPHANIE POON • *Princess Margaret Cancer Centre, University Health Network, Toronto, ON, Canada; Department of Medical Biophysics, University of Toronto, Toronto, ON, Canada*

PANAGIOTIS PRINOS • *Structural Genomics Consortium, University of Toronto, Toronto, ON, Canada*

SARATH RAMACHANDRAN • *Centre for Targeted Protein Degradation, Division of Biological Chemistry and Drug Discovery, School of Life Sciences, University of Dundee, James Black Centre, Dundee, UK*

VIVIANE REBER • *Institute of Molecular Systems Biology at ETH Zurich, Zurich, Switzerland*

MATTHEW B. ROBERS • *Promega Corporation, Madison, WI, USA*

ELISABETH M. ROTHWEILER • *Centre for Medicines Discovery, Nuffield Department of Medicine, University of Oxford, Oxford, UK; Target Discovery Institute, Nuffield Department of Medicine, University of Oxford, Oxford, UK*

AYSEGUL SAPMAZ • *Department of Cell and Chemical Biology, Leiden University Medical Centre, Leiden, The Netherlands*

ESPEN SCHALLMAYER • *Institute of Pharmaceutical Chemistry, Goethe University Frankfurt, Frankfurt, Germany*

STEPHAN SCHERPE • *Department of Cell and Chemical Biology, Leiden University Medical Centre, Leiden, The Netherlands*
ALEXANDRA STOLZ • *Institute of Biochemistry 2, Faculty of Medicine, Goethe University Frankfurt, Frankfurt am Main, Germany; Buchmann Institute for Molecular Life Sciences, Goethe University Frankfurt, Frankfurt am Main, Germany*
MAGDALENA M. SZEWCZYK • *Structural Genomics Consortium, University of Toronto, Toronto, ON, Canada*
LUTZ TAUTZ • *NCI-Designated Cancer Center, Sanford Burnham Prebys Medical Discovery Institute, La Jolla, CA, USA*
AMELIE TJADEN • *Institute of Pharmaceutical Chemistry, Goethe University Frankfurt, Frankfurt, Germany; Structural Genomics Consortium, BMLS, Goethe University Frankfurt, Frankfurt, Germany*
JAMES D. VASTA • *Promega Corporation, Madison, WI, USA*
VICTORIA VU • *Structural Genomics Consortium, University of Toronto, Toronto, ON, Canada*
CARROW I. WELLS • *Structural Genomics Consortium (SGC), UNC Eshelman School of Pharmacy, University of North Carolina at Chapel Hill (UNC-CH), Chapel Hill, NC, USA; Division of Chemical Biology and Medicinal Chemistry, UNC Eshelman School of Pharmacy, UNC-CH, Chapel Hill, NC, USA*
JENNIFER WILKINSON • *Promega Corporation, Madison, WI, USA*
JIAQIAN WU • *NCI-Designated Cancer Center, Sanford Burnham Prebys Medical Discovery Institute, La Jolla, CA, USA*

Chapter 1

An Introduction to Chemogenomics

Apirat Chaikuad and Daniel Merk

Abstract

Chemogenomics is an innovative approach in chemical biology that synergizes combinatorial chemistry and genomic and proteomic biology to systematically study the response of a biological system to a set of compounds, which can aid the identification and validation of biological targets as well as biologically active small-molecule agents responsible for a phenotypic outcome. Central to this strategy is a collection of chemically diverse compounds, a so-called chemogenomics library. Selection and annotation of vastly available chemogenomic compound candidates for an inclusion in such set present a challenge, but optimal compound selection is critical for success of chemogenomics. The library can be used in a wide variety of research applications from biological mechanism deconvolution to drug discovery. However, phenotypic screening methods are typically required to be high-throughput and equipped with a systematic analysis of complex biological–chemical interactions. This chapter provides a general outline to the chemogenomics approach, including concept and critical steps in all stages of this innovative chemical biology strategy.

Key words Chemogenomics, Chemical biology, Chemogenomic compound

1 Introduction: The Idea of Chemogenomics

Annotation of protein functions and importance in cellular processes has become a major challenge in the twenty-first-century post-genomic era. This not only helps illuminating our understanding of biology but may enable identification of key drivers of diseases that could ultimately accelerate discovery of new medicines. Many genetic approaches such as gene knock in or knock out have been used extensively and effectively in the past, yet elaborated experiments and unprecedented effects from an alteration of genome and proteome present a pitfall and may complicate elucidation of the biology of the target of interest. Chemical biology has emerged as an alternative through the use of chemical entities to modulate the activity of a target of interest. This strategy interferes less with cellular integrity and presents a model resembling drug

Daniel Merk and Apirat Chaikuad (eds.), *Chemogenomics: Methods and Protocols*, Methods in Molecular Biology, vol. 2706, https://doi.org/10.1007/978-1-0716-3397-7_1,

treatments. Despite relatively quicker and easier experimental procedures, there remain complications involving the choices of chemical tools. Potent and selective inhibitors, such as chemical probes [1, 2], are unarguably preferred; however, despite decades of research, these are hard to find and are not available for many classes of proteins, never mind those proteins that are hard to target by small molecules. In fact, even for a particular protein, those "good" inhibitors have also varying levels of selectivity, chemical and physical properties, and behaviors both in vitro and in vivo, all of which could affect phenotypic outcomes. Thus, the choice of a chemical tool and its use in experiments can lead to biased or even false perception of the functions of a target protein [3].

Chemogenomics has emerged as a strategy in chemical biology that systematically studies the response of an intact biological system to a set of compounds. In this definition, this approach involves screening a vast collection of chemical tools for the simultaneous identification and validation of biological targets and mechanism deconvolution of specific phenotypes [4, 5]. In addition, it offers in reverse another dimension in the search of compounds that are capable of interacting with targets of interest, which may enable the examination of classes of biologically active compounds against functionally or structurally related proteins within or across families. Chemogenomics usually deals with a large collection of, if not more, hundreds or thousands of biologically active compounds that constitute a "library" which can be classified into two types: (i) a set of diverse chemical agents with diverse, unrelated protein targets and (ii) a collection of small-molecule modulators against a singly selected target or a group of related targets. The compound libraries an be employed for identification of targets and mechanisms responsible for phenotypic outcomes in "forward chemogenomics," or used in "reverse chemogenomics" aiming to search for active small-molecule modulators of a given validated target [5]. In both approaches, due to a large number of compounds, screening methods are typically high-throughput and are usually coupled with the systematic analysis of complex chemical–biological interactions.

Chemogenomics presents a modern field of chemical biology, bringing together combinatorial chemistry and genomic and proteomic biology. Phenotypic screening combined with target-based approaches sees its greatest impacts in accelerating drug discovery. The application of this strategy facilitates not only the identification of new therapeutic targets but potentially drug repurposing, predictive toxicology as well as the discovery of alternative chemotherapeutic modalities [6]. With more accessible compound libraries and affordable high-throughput phenotypic screening platforms, chemogenomics has also started to see its application in a wide variety of academic research [4, 7].

2 What Makes a Chemogenomics Compound

The chemogenomics approach involves screening of libraries of molecules with known bioactivity profiles, i.e., target space, in phenotypic assays aiming to elucidate the involvement of a gene product in (patho-)physiological processes and its correlation with a phenotype/disease to move rapidly from gene to drug [5]. This chemical biology strategy is meant to facilitate discovery, evaluation, and validation of targets and active chemical agents. Chemical tools with highest quality in potency and selectivity such as chemical probes allow probing the molecular function, (patho-) physiological roles, and safety of macromolecular targets to pave a way toward novel therapeutics, but these highest-standard small-molecule agents are rare and currently cover only a minor fraction of druggable proteins [3, 8]. Chemogenomics can close this gap to accelerate target identification and validation by using a broader set of compounds with lower potency and selectivity until more chemical probes become available [8]. To compensate for their lower quality, chemogenomics compounds have to be highly annotated and should not be considered as individual chemical tools but as a set with ideally non- or low-overlapping chemical features and selectivity profiles to efficiently enable identification of a single (or a set of) macromolecular target(s) responsible for effects observed in library screening [6]. Additionally, chemogenomics compounds must be suitable for the use in cellular (phenotypic) assays [6]; i.e., they must have confirmed cellular activities and target engagement. Based on these considerations, chemogenomics compounds too must comply with specific criteria, which usually differ from the quality standards of chemical probes. Although several chemogenomics libraries have been published, such measures can be subjective and less transparent. Nevertheless, a guideline for chemogenomics compound selection has been defined by some initiatives such as the EUbOPEN consortium (www.eubopen.org) with exemplary criteria for their chemogenomics framework shown in Table 1.

One of the most critical aspects of chemogenomics compounds is target coverage. Ideally, small-molecule agents that share an intended target should have high chemical diversity and nonoverlapping off-target activity profiles (Fig. 1) [8]. Such characteristics allow target identification and validation with good confidence for a phenotypic outcome. Since no compound can be profiled on all possible macromolecular targets, the chemical diversity criterion provides an additional level of confidence as it reduces the probability of unknown overlapping activity, i.e., on targets that have not been tested.

Medicinal chemistry efforts over the last decades have generated a large collection of compounds capable of modulating the activity of a multitude of macromolecular targets, most of which

Table 1
Exemplary criteria for chemogenomics compounds targeting protein kinases (*see* also www.eubopen.org)

General criteria	Freedom to operate—Available for research use without IP issues HPLC purity ≥ 95% (AUC), identity confirmed by ESI-MS Preferably five different ligand chemotypes per protein target with complementary selectivity profiles A set of compounds for one target has appropriate selectivity (family-specific guidance); more stringent selectivity applies for targets with less available ligand chemotypes Toxicity data determined by multiplex assay at appropriate concentration for later use (e.g., cell death/lysis or significant morphological changes; toxicity from on-target inhibition excluded) Activity data in liability panel available at appropriate concentration for later use Compounds are manually rated to flag unstable compounds/undesired structures (e.g., polyols, peroxides, quinones)
Protein kinase inhibitor-specific selectivity guidance	In vitro IC_{50} or $K_d \leq 100$ nM or cellular $IC_{50} \leq 1$ μM Selectivity: Desirable—Screened across >100 kinases with S (>90% inhibition)[1] ≤ 0.025 with *S*: (off-targets/untested targets), $S = 0.025$ from an assay panel of 400 targets is equivalent to 10 off-targets with >90% inhibition

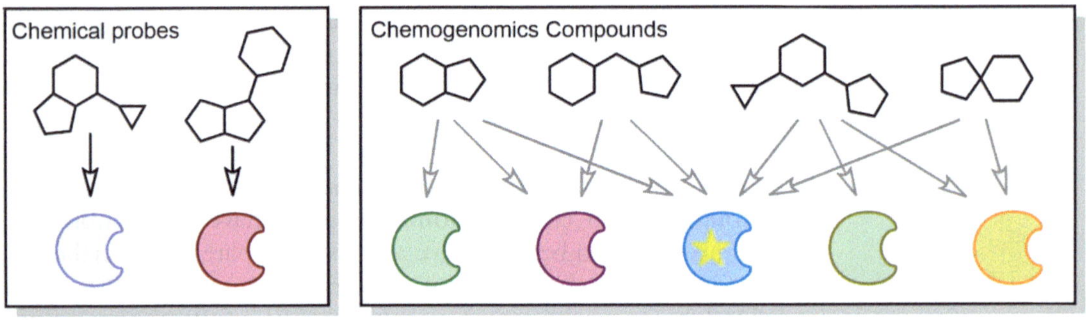

Fig. 1 Comparison of chemical probes and chemogenomics compounds [8]. Chemogenomics compounds usually have broader activity profiles and lower selectivity in comparison to chemical probes, thus they are typically used as a set in phenotypic screening. Inclusion of chemogenomics compounds with non-overlapping activity profiles and chemically diverse scaffolds enables target identification and mechanism deconvolution

can be considered chemogenomics compound candidates. Public compound/bioactivity databases like ChEMBL [9] which extract data from scientific publications provide a valuable asset for compilation of chemogenomics libraries [10]. Chapter 3 reports an example protocol for computational extraction and filtering of small-molecule candidates and associated data from public repositories. Considering the important characteristics of chemogenomics compounds, this protocol also outlines a selection strategy that promotes the important chemical diversity aspect in a library.

Like for all compounds intended to be used as a tool in biological assays, chemical quality control (QC), i.e., purity and identity, of chemogenomics compounds is deemed essential. Since chemogenomics typically deals with large numbers of compounds in a small quantity, an automated QC process that uses a small sample volume is desirable. In addition, due to substantially diverse chemical properties, the chemical QC methods must be applicable to compounds with a broad range of lipophilicity, molecular weight, salt form, and other characteristics. High-performance liquid chromatography (HPLC) coupled with UV detection for purity analysis and mass spectrometry (MS) for confirmation of identity fulfills these criteria and can be established with a high-throughput automation, exemplified by the LC-MS protocol in Chap. 4.

Nonspecific toxicity can affect virtually every biological assay and may cause serious misinterpretation [11, 12]. This is particularly relevant in chemogenomics as false positive outcomes owning to nonspecific toxicity in phenotypic screening may generate an invalid hypothesis and subsequently expensive follow-up experiments [11]. Thus, toxicity profiling forms a prerequisite before an inclusion of a compound in a chemogenomics set and should be performed at the concentration intended for the use in phenotypic screening, which depends on the potency and selectivity profile of each compound. Importantly, nonspecific toxicity must be carefully distinguished from target-mediated cell death to avoid exclusion of compounds, of which the mode-of-action causes cytotoxicity (e.g., inhibitors of BRD4 [13]). Like the chemical QC, toxicity testing needed for characterizing a large number of compound candidates requires a high level of automation. Chapters 5 and 6 describe high-throughput (multiplex) toxicity assays suitable for routine toxicity profiling of chemogenomics compounds.

Selectivity is another aspect that should be taken into consideration when assembling a compound library. In general, bioactive small molecules are often not exclusive to a target and tend to interact with structurally and/or functionally related proteins within a family ("off-targets"). The compounds that are to be included in a chemogenomics set should have a low number of off-targets and share little or no overlapping selectivity profiles. Selectivity may form a factor that determines the size of chemogenomics library: Considering the coverage of a target, only one highly specific compound with exclusive activity may be sufficient, while higher numbers are needed for less selective compounds. Sufficient within-family selectivity of the compounds in a set is critical to allow target identification and mechanism deconvolution of specific phenotypes with good confidence. Therefore, within-family selectivity profiling is very important and if possible should be performed using a uniform assay to allow standardized annotation. Examples of this process are detailed in Chaps. 7, 8 and 9.

Further examples include the annotation of selectivity of the chemogenomics libraries for protein kinases, the protein kinase inhibitor set (PKIS) library [7] and the kinase chemogenomic set (KCGS) [24], in which selectivity score (S) [14] and Gini coefficient [15, 16] were used.

Apart from within-family cross-reactivity, selectivity outside the main target protein family is another relevant factor; however, this cannot be comprehensively tested due to the vast number of potential off-targets. Nevertheless, a caution should be for small molecules with unspecific activity on a certain set of liablity targets, e.g., housekeeping proteins such as actin, BRD4, CDKs, and some important transcription factors, of which modulation tends to induce severe cytotoxicity or strong phenotypes. In principle, unless aiming for specific modulators of these proteins, such compounds should be excluded from the set. Activity profiling on a small number of such proteins (known also as liability or safety or promiscuity panel) [17, 18] is an approach to address this challenge. In drug discovery, early testing of unspecific adverse effects of candidates in a safety panel aims to reduce attrition in later development stages [18]. In the context of chemogenomics, a careful selection of targets in a liability panel can aid elimination of small molecules with unwanted phenotypic effects caused by activities on this class of proteins that would strongly compromise the successful use in phenotypic screening. In addition, a liability panel might consider other common off-targets that generally bind chemically diverse ligands, e.g., 5-HT_{2B} and PPARγ [17, 19, 20].

According to the similarity principle, chemically dissimilar compounds are considered less likely to have similar biological activity [21]. Among compound candidates that target the same protein of interest, it is therefore desirable to select a subset of compounds with the highest chemical diversity. Chemical diversity measured, for example, by fingerprints and by comparing scaffolds [22] should thus be considered in the final selection of compounds for a chemogenomics set from the collection of candidates with suitable features.

3 Chemogenomics Libraries

Assembling compound libraries is a major task of chemogenomics. Usually, defining target protein families or groups of functionally related proteins, e.g., GPCRs, kinases, nuclear receptors, ion channels, bromodomains, and other sets, is the first step, after which congeneric series of chemical entities capable of probing the actions of proteins in the selected specific classes are mined. As outlined in Subheading 2, stringent selection criteria should be adhered to for inclusion of compounds in the sets as consideration of these important features could determine the successful use of the libraries in phenotypic screening. Together with academic research efforts, several public–private partnership initiatives with an open-access policy have led to a large resource of biologically active small-molecule agents, of which information regarding properties such as targets, on-target activity, selectivity, cellular potency, and others can easily be obtained through numerous public databases such as ChEMBL, PubChem, Chemical Probes Portal, and Probe Miner. This has a benefit on the one hand in more choices of small-molecule modulators, yet on the other hand increases complexity of data mining and the compound selection process. Nevertheless, several bioinformatics and computational tools can aid this task. Examples of the development of a chemogenomics library can be found in [23, 24] and Chap. 2.

There are several chemogenomics libraries reported to date, which have different coverage of protein targets and families. The important information considering selection criteria and compound annotation are well described in some, while such details remain largely elusive in others. Thus, different compound sets have different merits and limitations in chemogenomics. In 2017, Jones and Bunnage [6] summarized examples of the libraries, including those from pharmaceutical industries such as the Pfizer chemogenomics library and the GlaxoSmithKline Biologically Diverse Compound Set (BDCS) and those that are available for public screening programs like the Sigma-Aldrich Library of Pharmacologically Active Compounds ($LOPAC^{1280}$), the Mechanism Interrogation PlatE library (MIPE), Molecular Libraries Program Probes from NIH, and the Protein Kinase Inhibitor Set (PKIS) library from the SGC-GSK partnership [7]. Recently, more chemogenomics sets have been reported such as the kinase chemogenomic set (KCGS) [24], the Novartis Institutes for BioMedical Research (NIBR) MoA Box (NIBR MoA Box) [25], and the SGC Probes and Donated Chemical Probes collection [26].

4 How to Use Chemogenomics Compound Sets

Screening a chemogenomics library in phenotypic assays aims to reveal targets, mechanisms, or even active small-molecule agents responsible for a specific phenotype, which would aid drug discovery programs [6]. The fact that effects in such settings are the results of target modulation by chemogenomics compounds simultaneously supports druggability of the target; i.e., that its pharmacological modulation is feasible [6]. Identifying chemical starting matter for drug discovery or hits for drug repurposing is not the main focus of chemogenomics [6] but may be a side product.

As chemogenomics compounds typically do not provide the target specificity of a chemical probe, they cannot be used individually but are meant to be applied as a set. Thus, suitable phenotypic screening should be high-throughput and able to accommodate testing a large number of small-molecule agents. Selecting suitable compound screening concentrations is a critical aspect in the application of chemogenomics libraries. While sufficient concentrations for robust modulation of the main target are necessary, a caution is for too high concentrations that are prone to increase off-target-mediated effects [6]. To avoid false hits, appropriate concentrations should be set on a per-compound basis rather than a fixed one-for-all value and should be simultaneously defined during library assembly.

Target and mechanism deconvolution for phenotypic effects are another critical task in chemogenomics. This process relies not only heavily on compound annotation but also on systematic analysis of complex chemical–biological interactions that form another integral part of a chemogenomics workflow. Subsequent target validation should then be performed in follow-up experiments involving negative control compounds and possibly genetic approaches [6]. This process may involve "reverse chemogenomics" using a set of target-based compounds, which may provide further benefits in the identification of active small-molecule agents.

There are a number of successful chemogenomic applications in a wide variety of drug discovery and biological research reported in literature, which demonstrate the potential of this innovative chemical biology strategy in target discovery and new target-disease correlation hypotheses that open new therapeutic avenues [27–29]. Together with these, diverse (phenotypic) screening methods with high-throughput nature coupled with systematic analysis of complex biological data have been established and shown their suitability in chemogenomics. In this book, the protocols of some assays with broad purposes are described in Chaps. 10, 11, 12, 13, 14, 15, 16 and 17.

5 Conclusion

Identification and validation of novel therapeutic targets presents a challenge and usually a bottleneck in the development of new medicines. Chemogenomics is an innovative chemical biology approach that synergizes combinatorial chemistry and biology to serve this task and has seen its greatest impact in accelerating drug discovery. This strategy does not only rely on high-quality potent and selective chemical tools such as chemical probes typically used in a single-target-based approaches, but also rather intends to exploit a benefit from the vast availability of biologically active small-molecule agents by screening them for the simultaneous identification and validation of biological targets and mechanism deconvolution of specific phenotypes.

The successful application of chemogenomics depends on several factors, including the quality of the compound collection and phenotypic screening methods. Assembly of a chemogenomics compound library forms the first challenging task, which requires rational selection and good annotation of suitable small-molecule agents. Together, the applicability of high-throughput phenotypic assays and systematic data analysis must be taken into consideration to enable validity of target and mechanism deconvolution outcomes. With many available biologically active small-molecule agents and affordable high-throughput phenotypic screening platforms, chemogenomics has recently gained a lot of attention in the post-genomic era with broader application not only in drug discovery but a multitude of academic biology research such as functional annotation of proteins. This book combines a number of protocols covering several key steps in chemogenomics from establishing the compound library to many phenotypic assays of wide-ranging purposes in order to provide guidance in applying chemogenomics toward pioneering new biology and therapeutic modalities.

References

1. Blagg J, Workman P (2017) Cancer Cell 32:9–25
2. Garbaccio RM, Parmee ER (2016) Cell Chem Biol 23:10–17
3. Arrowsmith CH, Audia JE, Austin C, Baell J, Bennett J, Blagg J, Bountra C, Brennan PE, Brown PJ, Bunnage ME, Buser-Doepner C, Campbell RM, Carter AJ, Cohen P, Copeland RA, Cravatt B, Dahlin JL, Dhanak D, Edwards AM, Frederiksen M, Frye SV, Gray N, Grimshaw CE, Hepworth D, Howe T, Huber KVM, Jin J, Knapp S, Kotz JD, Kruger RG, Lowe D, Mader MM, Marsden B, Mueller-Fahrnow A, Müller S, O'Hagan RC, Overington JP, Owen DR, Rosenberg SH, Roth B, Roth B, Ross R, Schapira M, Schreiber SL, Shoichet B, Sundström M, Superti-Furga G, Taunton J, Toledo-Sherman L, Walpole C, Walters MA, Willson TM, Workman P, Young RN, Zuercher WJ (2015) Nat Chem Biol 11:536–541
4. Steven Zheng XFS, Chan TF (2002) Curr Issues Mol Biol 4:33–43
5. Bredel M, Jacoby E (2004) Nat Rev Genet 5: 262–275
6. Jones LH, Bunnage ME (2017) Nat Rev Drug Discov 16:285–296
7. Elkins JM, Fedele V, Szklarz M, Abdul Azeez KR, Salah E, Mikolajczyk J, Romanov S, Sepetov N, Huang XP, Roth BL, Al Haj Zen A, Fourches D, Muratov E, Tropsha A, Morris J, Teicher BA, Kunkel M, Polley E, Lackey KE, Atkinson FL, Overington JP, Bamborough P, Müller S, Price DJ, Willson TM, Drewry DH, Knapp S, Zuercher WJ (2016) Nat Biotechnol 34:95–103
8. Müller S, Ackloo S, Al Chawaf A, Al-Lazikani B, Antolin A, Baell JB, Beck H, Beedie S, Betz UAK, Bezerra GA, Brennan PE, Brown D, Brown PJ, Bullock AN, Carter AJ, Chaikuad A, Chaineau M, Ciulli A, Collins I, Dreher J, Drewry D, Edfeldt K, Edwards AM, Egner U, Frye SV, Fuchs SM, Hall MD, Hartung IV, Hillisch A, Hitchcock SH, Homan E, Kannan N, Kiefer JR, Knapp S, Kostic M, Kubicek S, Leach AR, Lindemann S, Marsden BD, Matsui H, Meier JL, Merk D, Michel M, Morgan MR, Mueller-Fahrnow A, Owen DR, Perry BG, Rosenberg SH, Saikatendu KS, Schapira M, Scholten C, Sharma S, Simeonov A, Sundström M, Superti-Furga G, Todd MH, Tredup C, Vedadi M, Von Delft F, Willson TM, Winter GE, Workman P, Arrowsmith CH (2022) RSC Med Chem 13:13–21
9. Mendez D, Gaulton A, Bento AP, Chambers J, De Veij M, Félix E, Magariños MP, Mosquera JF, Mutowo P, Nowotka M, Gordillo-Marañón M, Hunter F, Junco L, Mugumbate G, Rodriguez-Lopez M, Atkinson F, Bosc N, Radoux CJ, Segura-Cabrera A, Hersey A, Leach AR (2019) Nucleic Acids Res 47:D930–D940
10. Isigkeit L, Chaikuad A, Merk D (2022) Molecules 27:2513
11. Moffat JG, Vincent F, Lee JA, Eder J, Prunotto M (2017) Nat Rev Drug Discov 16:531–543

12. Tjaden A, Chaikuad A, Kowarz E, Marschalek R, Knapp S, Schröder M, Müller S (2022) Molecules 27:1439
13. Filippakopoulos P, Qi J, Picaud S, Shen Y, Smith WB, Fedorov O, Morse EM, Keates T, Hickman TT, Felletar I, Philpott M, Munro S, McKeown MR, Wang Y, Christie AL, West N, Cameron MJ, Schwartz B, Heightman TD, La Thangue N, French CA, Wiest O, Kung AL, Knapp S, Bradner JE (2010) Nature 468: 1067–1073
14. Bosc N, Meyer C, Bonnet P (2017) BMC Bioinf 18:1–12
15. Ursu A, Childs-Disney JL, Angelbello AJ, Costales MG, Meyer SM, Disney MD (2020) ACS Chem Biol 15:2031
16. Wright Muelas M, Mughal F, O'Hagan S, Day PJ, Kell DB (2019) Sci Rep 2019 91:1–21
17. Sameshima T, Yukawa T, Hirozane Y, Yoshikawa M, Katoh T, Hara H, Yogo T, Miyahisa I, Okuda T, Miyamoto M, Naven R (2020) Chem Res Toxicol 33:154–161
18. Bendels S, Bissantz C, Fasching B, Gerebtzoff G, Guba W, Kansy M, Migeon J, Mohr S, Peters JU, Tillier F, Wyler R, Lerner C, Kramer C, Richter H, Roberts S (2019) J Pharmacol Toxicol Methods 99: 106609
19. Helmstädter M, Schierle S, Isigkeit L, Proschak E, Marschner JA, Merk D (2022) Int J Mol Sci 23:10070
20. Peters JU, Hert J, Bissantz C, Hillebrecht A, Gerebtzoff G, Bendels S, Tillier F, Migeon J, Fischer H, Guba W, Kansy M (2012) Drug Discov Today 17:325–335
21. Martin YC, Kofron JL, Traphagen LM (2002) J Med Chem 45:4350–4358
22. Nikolova N, Jaworska J (2003) QSAR Comb Sci 22:1006–1026
23. Dafniet B, Cerisier N, Boezio B, Clary A, Ducrot P, Dorval T, Gohier A, Brown D, Audouze K, Taboureau O (2021) J Cheminform 13:91
24. Wells CI, Al-Ali H, Andrews DM, Asquith CRM, Axtman AD, Dikic I, Ebner D, Ettmayer P, Fischer C, Frederiksen M, Futrell RE, Gray NS, Hatch SB, Knapp S, Lücking U, Michaelides M, Mills CE, Müller S, Owen D, Picado A, Saikatendu KS, Schröder M, Stolz A, Tellechea M, Turunen BJ, Vilar S, Wang J, Zuercher WJ, Willson TM, Drewry DH (2021) Int J Mol Sci 22:1–18
25. Canham SM, Wang Y, Cornett A, Auld DS, Baeschlin DK, Patoor M, Skaanderup PR, Honda A, Llamas L, Wendel G, Mapa FA, Aspesi P, Labbé-Giguère N, Gamber GG, Palacios DS, Schuffenhauer A, Deng Z, Nigsch F, Frederiksen M, Bushell SM, Rothman D, Jain RK, Hemmerle H, Briner K, Porter JA, Tallarico JA, Jenkins JL (2020) Cell Chem Biol 27: 1124–1129
26. Müller S, Ackloo S, Arrowsmith CH, Bauser M, Baryza JL, Blagg J, Böttcher J, Bountra C, Brown PJ, Bunnage ME, Carter AJ, Damerell D, Dötsch V, Drewry DH, Edwards AM, Edwards J, Elkins JM, Fischer C, Frye SV, Gollner A, Grimshaw CE, Ijzerman A, Hanke T, Hartung IV, Hitchcock S, Howe T, Hughes TV, Laufer S, Li VM, Liras S, Marsden BD, Matsui H, Mathias J, O'hagan RC, Owen DR, Pande V, Rauh D, Rosenberg SH, Roth BL, Schneider NS, Scholten C, Saikatendu KS, Simeonov A, Takizawa M, Tse C, Thompson PR, Treiber DK, Viana AY, Wells CI, Willson TM, Zuercher WJ, Knapp S, Mueller-Fahrnow A (2018) elife 7:e34311
27. Quintavalle M, Elia L, Price JH, Heynen-Genel S, Courtneidge SA (2011) Sci Signal 4: ra49
28. Schulz MMP, Reisen F, Zgraggen S, Fischer S, Yuen D, Kang GJ, Chen L, Schneider G, Detmar M (2012) Proc Natl Acad Sci U S A 109: E2665–E2674
29. Scheipl S, Barnard M, Cottone L, Jorgensen M, Drewry DH, Zuercher WJ, Turlais F, Ye H, Leite AP, Smith JA, Leithner A, Möller P, Brüderlein S, Guppy N, Amary F, Tirabosco R, Strauss SJ, Pillay N, Flanagan AM (2016) J Pathol 239:320–334

Chapter 2

Developing a Kinase Chemogenomic Set: Facilitating Investigation into Kinase Biology by Linking Phenotypes to Targets

Carrow I. Wells and David H. Drewry

Abstract

Advances in increasingly complex phenotypic screening with lower throughput have necessitated the screening of smaller more highly annotated sets. One such collection of compounds which has been recently assembled is the kinase chemogenomic set. This is a set of curated kinase inhibitors built upon previous iterations, PKIS and PKIS2, and donations from our partners. Each compound in the set has been carefully selected based on selectivity, potency, and kinome coverage. These compounds as a set have been made available to the scientific community, enabling phenotypic screens to identify kinases that drive novel biology. Additionally, the associated data deposited in the public domain have also been used to inform new inhibitor design. Further expansion of this set to complete kinome coverage will allow for a greater understanding of kinase biology and its role in disease.

Key words Kinase, Chemogenomics, Screening, Probes, Inhibitors, Drug discovery

1 Introduction

In drug discovery, there has been a major shift toward the development of disease-relevant assays to identify novel mechanisms of action that provides the desired phenotype. With this rise of increasingly more complex phenotypic assays, the need for screening sets to be well-curated is more important than ever. The utilization of chemogenomic sets in disease-relevant phenotypic screens can be a powerful approach. A chemogenomic set is a collection of well-annotated and validated pharmacological reagents [1]. Hits that arise from the screening of this type of set allow for rapid determination of which targets are being modulated to elicit the observed effect. Some details on the construction of chemogenomic sets and their application have been published by groups such as Novartis and Pfizer [1, 2]. More companies are also providing access to high-quality pharmacological probes such as

Daniel Merk and Apirat Chaikuad (eds.), *Chemogenomics: Methods and Protocols*, Methods in Molecular Biology, vol. 2706, https://doi.org/10.1007/978-1-0716-3397-7_2,

Boehringer Ingelheim through the opnMe project [3] and the AstraZeneca open innovation project where scientists can access over 14,000 molecules annotated toward 1600 targets (https://openinnovation.astrazeneca.com/). In addition to large pharmaceutical companies providing annotated sets, open science organizations such as the Structural Genomics Consortium (SGC) have also made it part of their mission to add to the growing number of annotated chemical probes available to researchers. They do this via several mechanisms, one being the donated probe program where pharmaceutical companies can donate potent and broadly characterized molecules to researchers to further elucidate the target biology [4]. These probes cover broad target space, but it is recognized that having sets focused on specific types of druggable targets in the proteome would also be beneficial. One such set of chemical tools from the SGC targets kinases.

There are over 500 human kinases, and they continue to be one of the most important drug target classes for the treatment of disease [5]. There are more than 70 FDA-approved kinase inhibitors (http://www.brimr.org/PKI/PKIs.htm) with many of the same targets being pursued by multiple companies (Fig. 1). Interestingly, among approved inhibitors from the last 5 years, only three of the approvals targeted novel kinases (SYK, TRKA/B/C, ROCK1/2), while twenty-eight approvals were for previously drugged kinases. This highlights the need for novel kinase-

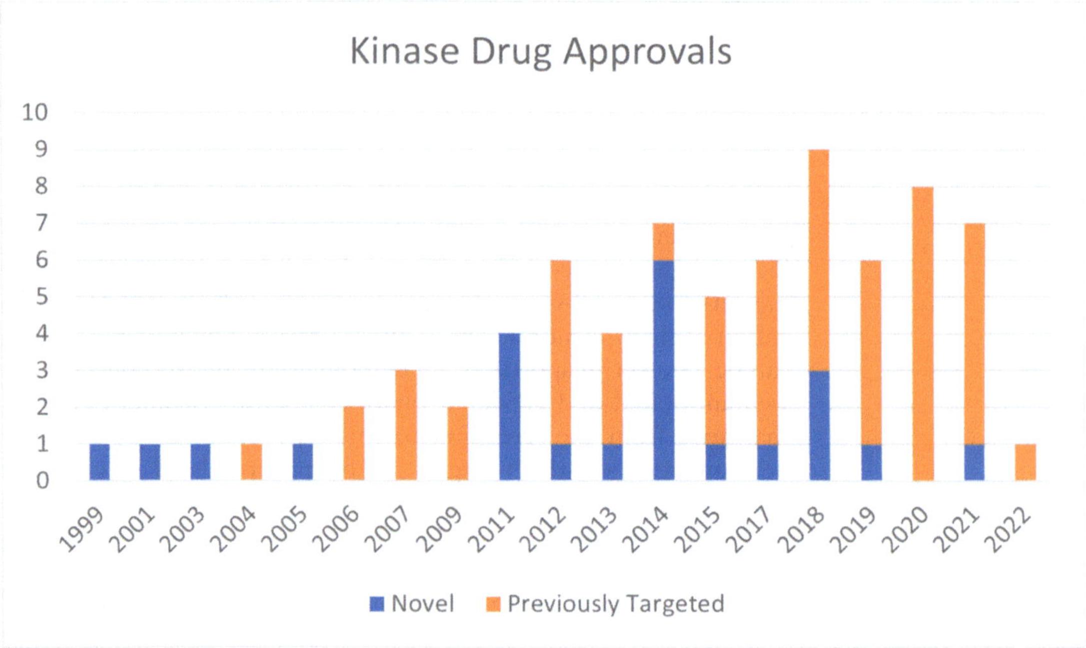

Fig. 1 All approved kinase inhibitors annotated for inhibition of a novel kinase target or for a kinase that has previously approved inhibitors

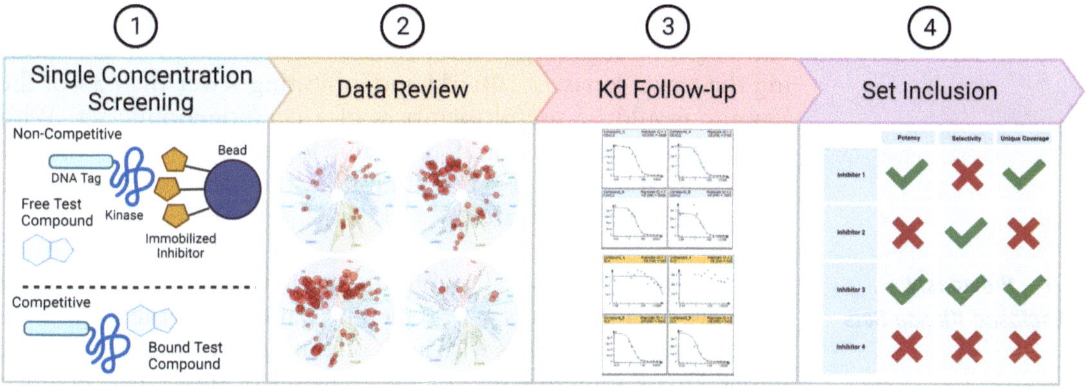

Fig. 2 Process to include an inhibitor in the set involves broad kinase profiling (e.g., KINOMEscan screening, step 1) and assessment of selectivity, potency, and the unique kinase coverage of each compound (step 4)

targeting probes to enable research into the rest of this promising drug class. The Structural Genomics Consortium (SGC) realized that this important drug class would benefit from a publicly available, well-annotated set to enable the scientific community to connect kinase inhibition to modulation of disease phenotypes. The kinase chemogenomic set (KCGS) is designed to provide the tools for the biological studies that set the stage for high-impact medicinal chemistry programs inside pharmaceutical and biotech companies targeting new kinases that have historically received less attention. We collected a candidate set of inhibitors that were donated by SGC partner pharmaceutical companies and academic research groups. With this collection of candidate inhibitors in hand, we used the pipeline described in Fig. 2 to determine suitability for inclusion in KCGS. This pipeline evaluates compounds for potency and selectivity as well as predicted developability and then compares activity profiles to ensure broad coverage.

2 Developing the Kinase Chemogenomic Set

There are over one million kinase publications and over 5000 human kinase crystal structures deposited [6]. In principle, this means that there is no shortage of inhibitors from which to choose to establish a kinase chemogenomic set. Unfortunately, only a fraction of the compounds that have been disclosed in papers and patents have been profiled against the kinome and an even smaller number of those meet potency and selectivity requirements to be included in a chemogenomic set. For a compound to be included in the set, it does not need to be exquisitely selective, and it just needs to be modestly selective and possess a complementary inhibition profile to other kinase inhibitors included in the set. Thus, these compounds do not need to be what the community refers to as chemical probes. Probes are small molecules characterized by high selectivity and potency which allow researchers to interrogate specific target biology [7]. It was determined that compounds inhibiting the target kinase <100 nM and inhibiting fewer than 5% of the kinome would be useful members of a chemogenomic set. With these simple criteria in mind, it became logical to conclude that we needed to start from a set of inhibitors that covers broad kinase space and is well-annotated.

2.1 Previously Disclosed Kinase Sets

There have been several publications that have disclosed large numbers of compounds with screening panel data (Table 1).

One set of inhibitors with such screening data available was the public kinase inhibitor set (PKIS). This set of molecules included 367 inhibitors published by GlaxoSmithKline. PKIS was profiled against 224 kinases at two concentrations and was made available to the scientific community for screening. This set of 364 compounds was originally synthesized to target twenty-four kinases and broad screening demonstrated >50% inhibition of 156 different kinases in the panel (by at least one compound) at a concentration of 1 μM. This set of molecules along with the annotation was then made available for free to the scientific community. The rapid uptake of the set by the scientific community and the resulting breadth of publications using PKIS compounds demonstrated the value of having such a set available to a vast array of researchers [16, 17]. Based on this success and the observed benefit to a variety of projects, we wanted to further expand the set to include more kinase coverage and an increased kinase inhibitor chemical space. To accomplish this, additional inhibitors were solicited from SGC

Table 1
Selection of large kinase inhibitor set papers that provide selectivity information across a range of kinase assays

Company/lab	Number of inhibitors	Number of kinases screened	Screening conc/ATP conc	Reference
SGC—PKIS2	645	442	1 μM with Kd follow-up/n.a	[8]
GSK—PKIS	367	224	100 nM and 1 μM/K_m ATP	[9]
GSK	577	203	10 μM/n.a	[10]
DiscoverX (Eurofins)	72	442	10 μM with Kd follow-up/n.a	[11]
Reaction biology Corp/ fox Chase Cancer Center	178	300	0.5 μM/10 μM	[12]
Abbott (Abbvie)	3800	172	pKi/5 μM–1 mM	[13]
EMD Millipore	158	234	1 μM and 10 μM/K_m of ATP	[14]
Kuster	243	253	3 nM–30 μM/endogenous	[15]

pharmaceutical partners to complement the GSK set. GSK further donated 645 inhibitors, some with almost unknown annotations, representing eighty-six diverse chemotypes. Additional inhibitors were donated to the SGC from Boehringer Ingelheim, Pfizer, Takeda, AbbVie, MSD, AstraZeneca, and Bayer. These compounds, although covering a broader chemical space, were also mostly unannotated for their kinase inhibition activities which necessitated screening to understand their potency and selectivity.

2.2 Kinase Activity Determination of KCGS

As previously mentioned, there are many good kinase tool molecules in the literature; however, many of these have been profiled in different assay formats at multiple vendors at various concentrations and consistent follow-up was not performed. To put our compounds all in the same context, all compounds being evaluated for inclusion in KCGS were profiled in the largest assay panel available, and all screening was performed at the same concentration with the same K_d follow-up criteria. The largest kinase panel commercially available at the time was the KINOME*scan* panel offered by Eurofins (formerly DiscoverX). This screen is an ATP-free competition binding assay that measures the ability of a free inhibitor to outcompete an immobilized inhibitor bound to a kinase of interest encoded with a DNA tag used to identify hits with a qPCR readout [11]. This panel provides percent binding in single concentration screening for 403 wild-type human kinases as well as some disease-relevant mutations of well-studied clinically relevant kinases (e.g., KIT, EGFR, ABL). All donated compounds were screened in this panel at the same concentration (1 μM) to ensure that all compounds were being evaluated similarly for both potency and selectivity. All compounds that had an $S_{10} \leq 0.05$ and an inhibition of >80% were followed up with K_d generation. For the compound–kinase pair to be included in the set, it needed a K_d of 100 nM or lower and had to inhibit 5% or less of the kinome at 90% inhibition at 1 μM treatment. In addition to these selectivity and potency requirements, the compound should provide some unique coverage or have complementary coverage to another inhibitor. For example, the compound SGC-AAK1–1 was included for its potency against AAK1, BMP2K, and RIOK1 and then another structurally related compound was also included that did not show binding to RIOK1 in the KINOME*scan* assay.

2.3 Size of Set

Our ideal size of a complete kinase chemogenomic set would be around 1000 inhibitors to cover the 400 screenable kinases. This would allow for multiple inhibitors per kinase (ideally 3) with the molecules covering the kinase having different chemotypes. The set is predominantly comprised of type I and II inhibitors, but of course allosteric compounds would be advantageous to include because of their enhanced selectivity. Currently, no PROTACs are included in the set; however, molecules such as these would be beneficial to validate the target of interest in downstream experiments. There is no need to include bespoke negative controls since those are currently built in because most compounds in the set are inactive against much of the kinome. Although we do not include specific negative controls in our set, they are still extremely useful molecules. They are typically very structurally related to a probe but do not inhibit the same kinase(s) [7]. This lack of on-target efficacy (compared to a probe) allows researchers to be able to directly attribute the cellular effects observed to the target of interest. If, however, a probe molecule is a hit when screening KCGS, the SGC does have specific negative control molecules that can be provided separately to help validate specific kinases such as AAK1, GAK, CSNK2A, and STK17B [18–21]. The current set that is being distributed is comprised of 187 compounds covering 215 kinases [22], but recently new 110 new compounds have been added that expand kinome coverage.

2.4 Chemotypes

To annotate each compound more fully, we wanted to have an automated way to assign each compound to a chemotype [22]. We did this first by defining common hinge binders using SMARTS and feeding the structures into the model to identify which hinge binder is present in each molecule. From this, we identified 119 possible chemotypes. Of these 119 possible chemotypes we defined, currently only 67 are represented in the set. Some of this difference in number of chemotype bins versus coverage is due to the inclusion of very similar bins. For example, 4-anilino-thieno[3,2]pyrimidine is covered, yet the very structurally related 4-anilino-thieno[2,3]pyrimidine is not in the set. There are also chemotypes that are overrepresented in the set, such as the 6-Ar-indazole and oxindole chemotypes with 16 and 13 exemplars, respectively. The sixteen 6-Ar-indazole kinase inhibitors cover sixteen kinases including kinases unique to the chemotype such as PDPK1 and AAK1. Interestingly, if you map the sixteen kinases covered by this chemotype, the kinases span six subfamilies but also inhibit highly related kinase pairs such as GSK3A and GSK3B and STK17A and STK17B (Fig. 3a). Furthermore, to demonstrate the importance of this chemotype, the examples we included do not inhibit kinases in the heavily studied TK and TKL branches of the kinome. The oxindole series has thirteen exemplars which were initially thought to only inhibit 30 kinases. Several of these compounds originated from the PKIS set, while other exemplars were in the PKIS expansion. These eight original compounds when profiled at Nanosyn were quite selective, but when later profiled in the KINOME*scan* panel, several of these compounds were discovered to be much less selective than previously anticipated. Updated kinome coverage is 66 kinases with several related kinases being identified (Fig. 3b). This highlights the caveats of screening such large numbers of compounds and using data from different data

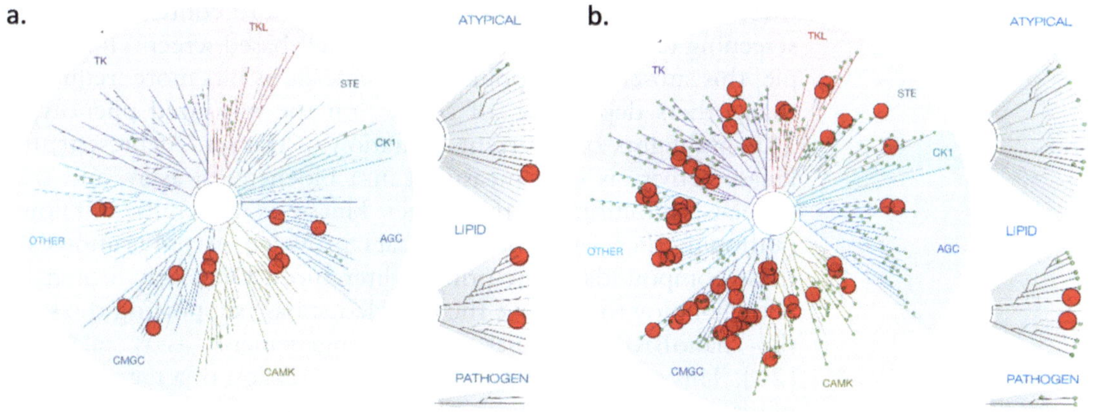

Fig. 3 Kinome coverage of selective and potent (S_{10} 1 μM <0.05, Kd <100 nM) compounds from two chemotypes (**a**) Sixteen 6-Ar-indazole and (**b**) Thirteen oxindole inhibitors

sources. The screening results should always be looked upon as preliminary data to point the researcher in certain directions that necessitate careful follow-up to reveal the target(s) driving the observed biology. Different assay platforms will sometimes lead to different results (even for the same kinase) due to differences in assay format, assay conditions, or the protein itself (e.g., different constructs may be used).

2.5 Library Properties

In assembling a set to be utilized in cell-based experiments, we also wanted to ensure that the compounds had reasonable developability properties such as solubility and logP. Most of our compounds fell in the moderately soluble to soluble range [22]. However, about 10% were predicted to be poorly soluble. These 22 predicted poorly soluble compounds include published tools like ERK5-IN-1 which is a potent and selective inhibitor of MAPK7 (ERK5), and few other compounds are known to meet KCGS criteria and inhibit this kinase [23]. The ideal logP (or clogP) is generally considered an important factor to bear in mind during drug discovery efforts [24]. The majority of KCGS conforms to clogP <5; in fact, there are only four compounds with a clogP >5. All of these are, as one might expect, also predicted to be poorly soluble. In future iterations, we aim to remove these poorly soluble compounds are replace them with molecules with more desirable developability properties. In addition to calculated properties, we wanted to ensure that no compounds in the set were broadly cytotoxic. To test this, the set was evaluated in 15 cancer cell lines and 2 control cell lines to ensure no compound broadly caused cell death [25]. Fifteen compounds were found to affect a number of the cell lines screened, but these compounds shared common targets, PLKs, and the lipid kinases PIK3C and PI4KB, so the toxicity was not due to intrinsic properties of the compounds. However, having these data available is useful to provide more context as further screening data are collected. In cancer cell-based screens, for example, this subset of compounds may emerge as hits more frequently than others depending, of course, on the measured phenotype. This highlights one current limitation of the set. At present, the kinase activity is only measured in a biochemical assay, and true cellular inhibition of the target kinase is generally unknown (although there may be such data provided in publications on these compounds in the primary literature). One assay format we could employ to improve the knowledge base supporting the set is the nanoBRET cellular target engagement assay platform [26]. This system allows for the overexpression of a tagged kinase of interest and IC_{50} determination in the cell based on competition with an ATP competitive tracer from the active site. This would provide an additional layer of activity data bridging the biochemical assay data that have already been collected with the phenotypic assay output from collaborators.

2.6 Kinase Coverage

Currently, in the KCGS, we cover 215 kinases. This represents a little more than half of the screenable kinome. We define screenable kinome here as those kinases that have a commercially available assay [8]. Some compounds in KCGS are donations from Nathanael Gray's laboratory, a laboratory that has published many high-quality kinase probes for kinases such as CDK7 and FGFR [27, 28]. Other compounds have come from donations from SGC partners such as Pfizer, Boehringer Ingelheim, and Takeda, and the set will be expanded to include Abbvie, Bayer, Merck, and AstraZeneca donated compounds. Most of the compounds in the set come from GSK (Fig. 4). This is mainly due to the donation of

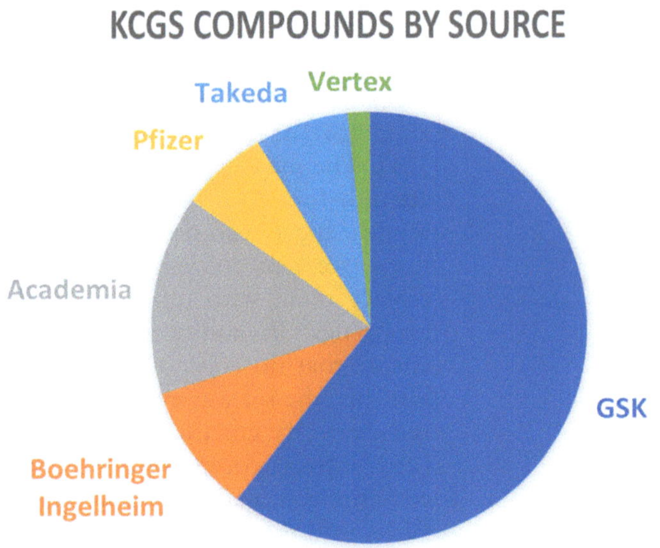

Fig. 4 Compounds included in KCGS sorted by original source

over 900 compounds in total to the SGC. For "uncovered" kinases, we are working to identify not only new KCGS eligible compounds but take it a step further to develop chemical probes for these targets. As an example, the kinases CSNK2a1 and CSNK2a2 were both uncovered kinases in KCGS. There is a published molecule, CX-4945, which inhibits both CSNK2a1 and CSNK2a2 (as well as 7 additional kinases out of the 238 that were evaluated) [29]. We have opted to only include compounds for which a partner has formally agreed that we can distribute the inhibitor, so CX-4945 was not included in the set. Instead, utilizing literature inhibition information on CSNK2a, as well as internal screening data, we designed and synthesized a dual CSNK2 probe that will be in the expanded KCGS [20]. While we have made considerable progress in covering the kinome with KCGS, we still have work to do. We anticipate that with continued effort we will be able to identify and select compounds that inhibit most of the rest of the screenable kinome. Such a completed set will fully enable this resource that facilitates matching specific phenotypes to all members of the screenable kinome.

2.7 Plating

To provide the set for medium throughput screening, and so the final set would fit in three plates, we decided to plate the set in 384-well format. The set was plated using an Echo Acoustic Dispenser to enable accurate dispensing of each compound. To do this, 12 μL of a 10 mM DMSO stock solution each inhibitor was added to an Echo-qualified low-dead-volume compatible source microplate. From each source plate, 11 KCGS compound plates could be stamped and sealed using a heat sealer. The remaining volume from all the source plates was combined to generate a quality control plate to ensure that the correct compound was in each well. Each well in a distributed KCGS plate contains 1 μL of inhibitor at a concentration of 10 mM stock in DMSO. This allows the receiving researcher to further dilute with 9 μL DMSO to obtain 10 μL of a 1 mM DMSO stock. From there, 1 μL can be used and diluted with 99 μL of assay medium to provide a 10 μM solution. With a 10 μM in DMSO stock solution available, one can deliver the compounds to the desired phenotypic assays in a standard manner. For example, the test inhibitors can be added as a tenth of the assay volume, so, for example, if the total assay volume is 100 μL, then one would add 10 μL of each inhibitor for the recommended final assay concentration of 1 μM. If assay throughput allows, this dilution protocol also provides plenty of material for multiple replicates and the ability to screen at lower concentrations to observe a dose response of hits. To facilitate the follow-up for hit validation experiments that are needed after initial screening, SGC-UNC will provide up to five additional compound aliquots (1 μL at 10 μM) of the hits from a primary screen.

2.8 Distribution

Our mission with this set was simple—to provide research compounds that are backed by high-quality data to enable further research into kinase inhibitors. To continue to screen and provide these compounds for years to come, we had to do something that we had never done before and that was charge for the set. Proceeds from this charge to date have allowed us to complete key experiments that enhance the quality of the set as well as contribute to the sustainability of the set. These experiments included additional screening at Eurofins DiscoverX with K_d follow-up and the resynthesis of key compounds at a contract research organization. We were also able to employ a full-time scientist for a year who was devoted to overseeing the resynthesis efforts, coordinating the collection of compounds, and the plating and distribution of the set. The set was initially directly distributed from our academic laboratory at the University of North Carolina but was then transitioned to distribution through cancertools.org. This set, much like our first two sets, PKIS and PKIS2, has seen rapid uptake by the scientific community. This uptake means the set will be utilized in a broad range of biological contexts. A primary reason for sharing compounds is that, through sharing, many more experiments can be performed than we could ever perform ourselves and we tap into the expertise and breadth and depth of knowledge of the global scientific community. Our sets have indeed been used in more biological experiments than we could ever do on our own, ranging from studies in chordoma, a rare cancer, to screens to identify leads for neglected tropical diseases.

3 How Researchers Have used PKIS and KCGS

We believe that creating and distributing highly annotated sets provides a value to the scientific community that is not attainable with only evaluating commercial molecules. Many molecules distributed commercially do not have full kinase annotation which makes ascribing phenotypic response to a specific target difficult. Screening KCGS as a set allows for a putative target list to be generated which can then be interrogated by additional experiments. A few examples of the use of our chemogenomic sets follow (Fig. 5). One example comes from a group of scientists interested in a rare skin disease disorder, epidermolysis bullosa simplex (EBS), for which there is no known treatment beyond palliative care [30]. This team screened PKIS to identify kinase inhibitors that reduced keratin aggregate formation [31]. Most of the inhibitors that showed activity in the assay were linked to the EGFR signaling pathway. These results were further validated by evaluation of the clinically approved EGFR inhibitor Afatinib. Afatinib also drove EBS cells into a more quiescent state, and thus, they were to quickly align the target and the phenotype [31].

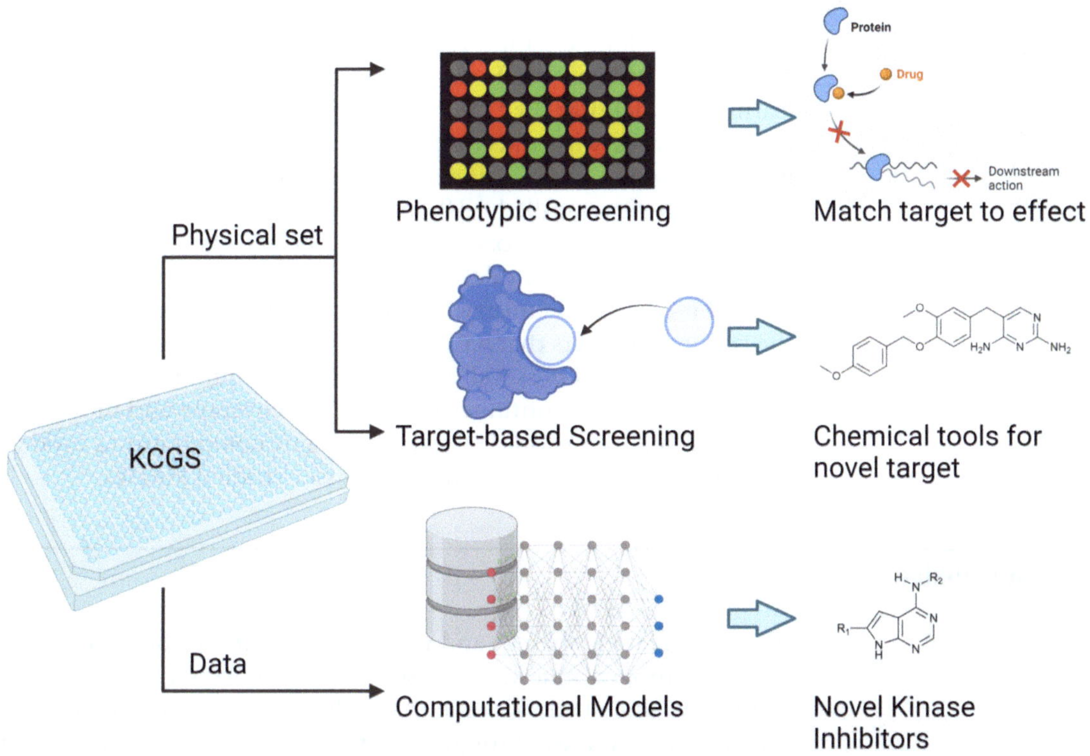

Fig. 5 Depiction of how collaborators are using both the physical KCGS set and the data that have been made publicly available to match a target to a phenotype, find tools for new targets, and use models to develop novel kinase inhibitors

There are also kinases for which there have been no cell-active inhibitors reported, so screening a collection of high-quality narrow-spectrum inhibitors could identify a reasonably selective chemical starting point. One such target is ERK3, a kinase which has been implicated in cancer invasiveness and migration but with no reported inhibitors to evaluate the implications of cellular inhibition. Our collaborators were able to screen our PKIS set and identify several cell-active inhibitors of ERK3 [32]. These compounds can now be taken into further studies to understand the impact of ERK inhibitors in relevant cell lines and used as starting points for further medicinal chemistry optimization into selective chemical probes.

In addition to providing physical samples of the compounds, releasing all the profiling data into the public domain has value as well and allows groups to generate SAR hypotheses to then synthesize molecules to test those hypotheses. One such example is a report from Henderson et al. who explored the published data around the pyrazolopyridazine series and from that data identified activity cliffs to predict molecules that were DYRK1a selective over GSK3 and CDKs [33]. Other groups have also been able to use the data to benefit their research by building models that can predict activities for kinases. One recent example comes from Bao et al. This group utilized PKIS and PKIS2 data to generate a new deep learning model that they call "an auxiliary multitask graph isomorphism network with uncertainly weighing." This multitask graph neural network (GNN) model they developed, available via a free online web server called KIP, may prove useful for prediction of activities of new inhibitors. [34] It will be fascinating, and important, to "test drive" models such as this, in the design-make-test cycle of medicinal chemistry. These examples highlight just a few of the research projects that have been enabled by the distribution and publication of KCGS as well as previous sets of inhibitors (PKIS and PKIS2). As the set continues to expand in coverage and more data are published, this should help further refine computational models and enable further research into the lesser-studied kinases.

4 Conclusion

We have created, shared, and published the first publicly available kinase chemogenomic set. This set currently covers a bit more than half of the screenable kinome. As such, we are actively looking to expand our coverage through donations from pharmaceutical and academic collaborators combined with internal medicinal chemistry projects focused on inhibitors for more members of the understudied kinome. In addition to our goal of expanded coverage, we realize that the value of the set can be increased by further annotation of the set with cell-based inhibition data. Cellular data, such as could be provided by the NanoBRET in cell target engagement assay, will help bridge the gap between biochemical assays and phenotypic results [26, 35]. We acknowledge that the current version of KCGS is simply a step along a path to full kinome coverage, and full kinome coverage will be a step toward eventual full proteome coverage. The SGC is fully embracing this journey and is hard at work to galvanize the scientific community behind an initiative called Target 2035, with the goal of identification of a modulator for every protein in the proteome. This is an ambitious and important goal that will require the development of new strategies and novel technologies and will also require global engagement, organization, and cooperation. Without such efforts though, the secrets of much of the proteome will remain hidden, and the full promise of the sequencing of the human genome will not be realized.

Acknowledgement

This work was supported by the Eshelman Institute for Innovation (http://unceii.org/). The SGC is a registered charity (number 1097737) that receives funds from AbbVie, Bayer Pharma AG, Boehringer Ingelheim, Canada Foundation for Innovation, Eshelman Institute for Innovation, Genome Canada, Innovative Medicines Initiative (EU/EFPIA) [ULTRA-DD grant no. 115766], Janssen, Merck & Co., Novartis Pharma AG, Ontario Ministry of Economic Development and Innovation, Pfizer, São Paulo Research Foundation-FAPESP, Takeda, and Wellcome Trust [092809/Z/10/Z]. GlaxoSmithKline, Pfizer, Takeda, AstraZeneca, Boehringer Ingelheim, Bayer, Abbvie, Merck, and Nathanael Gray are gratefully acknowledged for donation of KCGS compounds.

References

1. Jones LH, Bunnage ME (2017) Applications of chemogenomic library screening in drug discovery. Nat Rev Drug Discov 16(4): 285–296
2. Canham SM, Wang Y, Cornett A, Auld DS, Baeschlin DK, Patoor M et al (2020) Systematic Chemogenetic library assembly. Cell Chem Biol 27(9):1124–1129
3. Gollner A, Koster M, Nicklin P, Trieselmann T, Klein E, Vlach J et al (2022) opnMe.com: a digital initiative for sharing tools with the biomedical research community. Nat Rev Drug Discov 21(7):475–476
4. Müller S, Ackloo S, Arrowsmith CH, Bauser M, Baryza JL, Blagg J et al (2018) Science Forum: donated chemical probes for open science. Elife 7:e34311
5. Manning G, Whyte DB, Martinez R, Hunter T, Sudarsanam S (2002) The protein kinase complement of the human genome. Science 298(5600):1912–1934
6. Wu P, Nielsen TE, Clausen MH (2015) FDA-approved small-molecule kinase inhibitors. Trends Pharmacol Sci 36(7):422–439
7. Arrowsmith CH, Audia JE, Austin C, Baell J, Bennett J, Blagg J et al (2015) The promise and peril of chemical probes. Nat Chem Biol 11(8):536–541
8. Drewry DH, Wells CI, Andrews DM, Angell R, Al-Ali H, Axtman AD et al (2017) Progress towards a public chemogenomic set for protein kinases and a call for contributions. PLoS One 12(8):e0181585
9. Elkins JM, Fedele V, Szklarz M, Abdul Azeez KR, Salah E, Mikolajczyk J et al (2016) Comprehensive characterization of the published kinase inhibitor set. Nat Biotechnol 34(1):95–103
10. Bamborough P, Drewry D, Harper G, Smith GK, Schneider K (2008) Assessment of chemical coverage of kinome space and its implications for kinase drug discovery. J Med Chem 51(24):7898–7914
11. Davis MI, Hunt JP, Herrgard S, Ciceri P, Wodicka LM, Pallares G et al (2011) Comprehensive analysis of kinase inhibitor selectivity. Nat Biotechnol 29(11):1046–1051
12. Anastassiadis T, Deacon SW, Devarajan K, Ma H, Peterson JR (2011) Comprehensive assay of kinase catalytic activity reveals features of kinase inhibitor selectivity. Nat Biotechnol 29(11):1039–1045
13. Metz JT, Johnson EF, Soni NB, Merta PJ, Kifle L, Hajduk PJ (2011) Navigating the kinome. Nat Chem Biol 7(4):200–202
14. Gao Y, Davies SP, Augustin M, Woodward A, Patel UA, Kovelman R et al (2013) A broad activity screen in support of a chemogenomic map for kinase signalling research and drug discovery. Biochem J 451(2):313–328
15. Klaeger S, Heinzlmeir S, Wilhelm M, Polzer H, Vick B, Koenig PA et al (2017) The target landscape of clinical kinase drugs. Science 358(6367):eaan4368
16. Dranchak P, MacArthur R, Guha R, Zuercher WJ, Drewry DH, Auld DS et al (2013) Profile of the GSK published protein kinase inhibitor set across ATP-dependent and-independent luciferases: implications for reporter-gene assays. PLoS One 8(3):e57888

17. Khagi S, Sengupta S, Tabtieng T, Cochran BH (2015) Abstract C201: activity of the GlaxoSmithKline kinase inhibitor set on glioblastoma stem cells. Mol Cancer Ther 14-(12_Supplement_2):C201-C
18. Wells C, Counago RM, Limas JC, Almeida TL, Cook JG, Drewry DH et al (2020) SGC-AAK1-1: a chemical probe targeting AAK1 and BMP2K. ACS Med Chem Lett 11(3):340–345
19. Asquith CRM, Berger BT, Wan J, Bennett JM, Capuzzi SJ, Crona DJ et al (2019) SGC-GAK-1: a chemical probe for cyclin G associated kinase (GAK). J Med Chem 62(5):2830–2836
20. Wells CI, Drewry DH, Pickett JE, Tjaden A, Kramer A, Muller S et al (2021) Development of a potent and selective chemical probe for the pleiotropic kinase CK2. Cell. Chem Biol 28(4):546–58 e10
21. Picado A, Chaikuad A, Wells CI, Shrestha S, Zuercher WJ, Pickett JE et al (2020) A chemical probe for dark kinase STK17B derives its potency and high selectivity through a unique P-loop conformation. J Med Chem 63(23):14626–14646
22. Wells CI, Al-Ali H, Andrews DM, Asquith CRM, Axtman AD, Dikic I et al (2021) The kinase Chemogenomic set (KCGS): an open science resource for kinase vulnerability identification. Int J Mol Sci 22(2):566
23. Deng X, Elkins JM, Zhang J, Yang Q, Erazo T, Gomez N et al (2013) Structural determinants for ERK5 (MAPK7) and leucine rich repeat kinase 2 activities of benzo[e]pyrimido-[5,4-b]diazepine-6(11H)-ones. Eur J Med Chem 70:758–767
24. Lipinski CA, Lombardo F, Dominy BW, Feeney PJ (2001) Experimental and computational approaches to estimate solubility and permeability in drug discovery and development settings. Adv Drug Deliv Rev 46(1–3):3–26
25. Schirmer AU, Driver LM, Zhao MT, Wells CI, Pickett JE, O'Bryne SN et al (2022) Non-canonical role of Hippo tumor suppressor serine/threonine kinase 3 STK3 in prostate cancer. Mol Ther 30(1):485–500
26. Vasta JD, Corona CR, Wilkinson J, Zimprich CA, Hartnett JR, Ingold MR et al (2018) Quantitative, wide-spectrum kinase profiling in live cells for assessing the effect of cellular ATP on target engagement. Cell Chem Biol 25(2):206–14. e11
27. Kwiatkowski N, Zhang T, Rahl PB, Abraham BJ, Reddy J, Ficarro SB et al (2014) Targeting transcription regulation in cancer with a covalent CDK7 inhibitor. Nature 511(7511):616–620
28. Zhou W, Hur W, McDermott U, Dutt A, Xian W, Ficarro SB et al (2010) A structure-guided approach to creating covalent FGFR inhibitors. Chem Biol 17(3):285–295
29. Siddiqui-Jain A, Drygin D, Streiner N, Chua P, Pierre F, O'Brien SE et al (2010) CX-4945, an orally bioavailable selective inhibitor of protein kinase CK2, inhibits prosurvival and angiogenic signaling and exhibits antitumor efficacy. Cancer Res 70(24):10288–10298
30. Uitto J, McGrath JA, Rodeck U, Bruckner-Tuderman L, Robinson EC (2010) Progress in epidermolysis bullosa research: toward treatment and cure. J Invest Dermatol 130(7):1778–1784
31. Tan TS, Common JEA, Lim JSY, Badowski C, Firdaus MJ, Leonardi SS et al (2021) A cell-based drug discovery assay identifies inhibition of cell stress responses as a new approach to treatment of epidermolysis bullosa simplex. J Cell Sci 134(19):jcs258409
32. Schroder M, Filippakopoulos P, Schwalm MP, Ferrer CA, Drewry DH, Knapp S et al (2020) Crystal structure and inhibitor identifications reveal targeting opportunity for the atypical MAPK kinase ERK3. Int J Mol Sci 21(21):7953
33. Henderson SH, Sorrell F, Bennett J, Hanley MT, Robinson S, Hopkins Navratilova I et al (2020) Mining public domain data to develop selective DYRK1A inhibitors. ACS Med Chem Lett 11(8):1620–1626
34. Bao L, Wang Z, Wu Z, Luo H, Yu J, Kang Y et al (2023) Kinome-wide polypharmacology profiling of small molecules by multi-task graph isomorphism network approach. Acta Pharm Sin B 13(1):54–67
35. Wells CI, Vasta JD, Corona CR, Wilkinson J, Zimprich CA, Ingold MR et al (2020) Quantifying CDK inhibitor selectivity in live cells. Nat Commun 11(1):2743

Chapter 3

Compilation of Custom Compound/Bioactivity Datasets from Public Repositories

Laura Isigkeit and Daniel Merk

Abstract

Public repositories containing compound–bioactivity data for millions of small molecules offer a valuable resource for chemogenomic compound candidate search. Nonetheless, owning to nonuniform data mining, these databases are often incomplete, thus advocating the combined use of data from several repositories to increase target coverage and data accuracy. Here, we present a workflow to generate custom datasets from public databases for mining chemogenomic compound candidates. The compiled set provides flags for differences in structural and bioactivity data and enables rapid extraction of potent and selective bioactive compounds.

Key words ChEMBL, Public databases, Big data, Data curation, Compound selectivity

1 Introduction

Identification and validation of protein targets with potential therapeutic benefit are a critical step in the development of new medicines. In the absence of high-quality chemical probes for the majority of druggable proteins [1, 2], compounds with low level of selectivity can be employed in chemogenomics [3]. This chemical biology approach evaluates phenotypic effects of bioactive molecules in biological systems at a large scale and thus relies on a large set of well-characterized compounds with well-annotated biological activity [4, 5]. Medicinal chemistry efforts in academia and industry over the last decades have generated a vast number of modulators for diverse protein targets, and these form a valuable asset for chemogenomics. Public databases like PubChem [6], ChEMBL [7], and IUPHAR [8] gather the structures of bioactive molecules and their biological activities from scientific literature and make them accessible as a resource for the compilation of chemogenomic libraries.

Daniel Merk and Apirat Chaikuad (eds.), *Chemogenomics: Methods and Protocols*, Methods in Molecular Biology, vol. 2706, https://doi.org/10.1007/978-1-0716-3397-7_3,

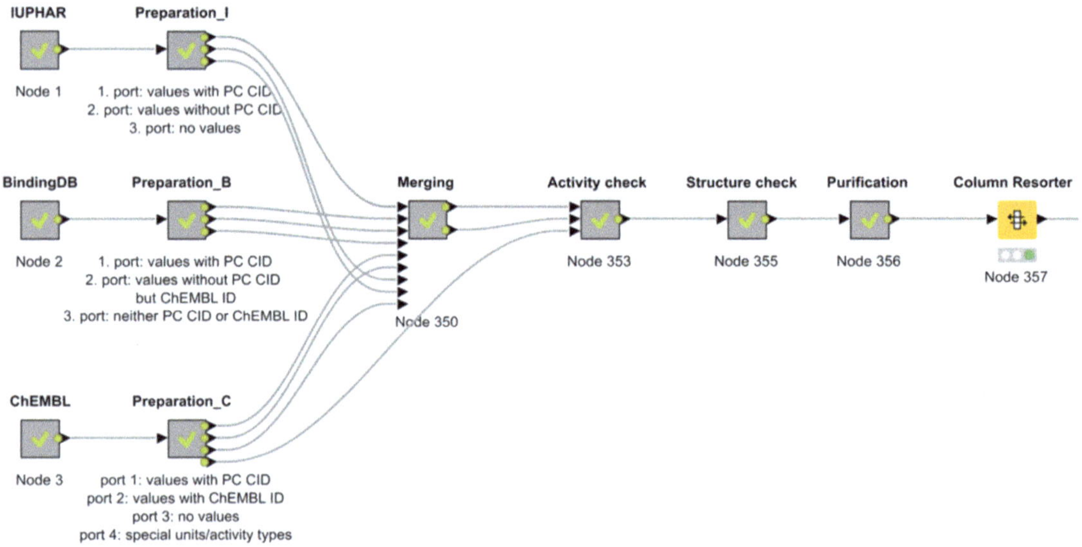

Fig. 1 Workflow steps overview to create a consensus compound/bioactivity dataset from the databases IUPHAR/BPS, BindingDB and ChEMBL. The workflow can be easily adapted to other databases

The merits of a chemogenomic library depend heavily on the selection and annotation of chemogenomic compounds [1, 3, 4, 9]. Key features of the compounds such as nonoverlapping selectivity profiles and chemical diversity are critical to enable target and/or mechanism deconvolution of phenotypic responses [1, 3, 4, 9]. Assembling a chemogenomic set from bioactive molecules reported in the literature is an elaborate task, which can benefit remarkably from the use of compound–bioactivity data in public repositories.

This chapter provides a workflow to generate a virtual database [10] of bioactive molecules incorporating also the associated bioactivity data from various public sources, which can be exploited for chemogenomic compound selection. As selectivity is a key aspect of chemogenomic compounds, this virtual database includes the comparison of bioactivity annotations collected from different sources to improve confidence. An overview of the workflow to assemble the dataset is shown in Fig. 1.

2 Materials

2.1 Computational Framework – Software, Programming Language, and Tools

1. Konstanz Information Miner (KNIME, available at https://www.knime.com/) with the KNIME Extensions: KNIME Expressions, RDKit Nodes Feature, KNIME Base Chemistry Types and Nodes, KNIME Parallel Chunk Loop Nodes, Vernalis KNIME Nodes, KNIME Python 2 Integration (legacy).
2. Python (e.g., version 3.7 available at Python Software Foundation, https://www.python.org).

3. Anaconda for Python (https://www.anaconda.com).
4. RDKit (https://www.rdkit.org/; https://www.rdkit.org/docs/Install.html#introduction-to-conda. The created environment has to be configured in KNIME *File* > *KNIME* > *Python (legacy)* as described in the KNIME Python Integration Installation Guide at https://docs.knime.com/2022-06/python_installation_guide/#_introduction).
5. PostgreSQL (https://www.postgresql.org/; needed to handle the ChEMBL database).

2.2 Databases

1. Approved protein symbols from HGNC database: Download all symbols from https://www.genenames.org/download/custom/ and select the following IDs: Approved symbol, Approved name, UniProt ID, Ensembl gene ID, Previous symbols, Previous name, Alias symbols; copy and paste the resulting output from the webpage in a text file and save it as HGNC_symbols.txt.
2. IUPHAR/BPS: Download the following CSV files from https://www.guidetopharmacology.org/download.jsp: ligand list (contains ligand information including structure as SMILES), interaction data (contains information on bioactivities), and target and family list (contains information on targets and target families). The human organism is declared as "human" in IUPHAR.
3. BindingDB: Download the following files from https://www.bindingdb.org/rwd/bind/chemsearch/marvin/SDFdownload.jsp?all_download=yes: BindingDB_All_2D_*.tsv.zip (ligand–target affinity datasets contain all molecules, bioactivities, and targets) and unzip the file, BindingDB_CID.txt (lists and identifier mapping contain PubChem CID). The human organism is declared as "homo sapiens" in BindingDB.
4. ChEMBL: Download the latest PostgreSQL-Dump (*_postgresql.tar.gz) version of ChEMBL from https://ftp.ebi.ac.uk/pub/databases/chembl/ChEMBLdb/latest/. Create a database in PostgreSQL, load the dump file, and use the ReadMe file of the downloaded dump file to recreate the database. Note the database name, hostname, port, username, and password to access to the created database via KNIME.
5. To merge all information about compounds, assays, and targets, you need the entity–relation diagram (ERD) from ChEMBL (https://ftp.ebi.ac.uk/pub/databases/chembl/ChEMBLdb/latest/ – chembl_XXX_schema.png). The human organism is declared as "homo sapiens" in ChEMBL.

3 Methods

The following protocol is formatted as follows: Bold text corresponds to the names of KNIME nodes, text in brackets describes settings of the corresponding node, and <> denotes patterns or delimiters. Do not include the <> in the KNIME node. Use default settings when no settings are specified. The symbol “\n” in node settings means the following expression should be in a new line. The names of metanodes are written in italics. KNIME provides useful explanations of the required KNIME nodes under the Description tab. Figure 1 gives an overview of the workflow to create a consensus compound/bioactivity dataset.

3.1 Data Preparation

3.1.1 HGNC File

1. Load the HGNC_symbols.txt (*see* Subheading 2.2, **step 1**) file into KNIME using **File Reader** and convert all symbols to lowercase using **String Manipulation (Multi Column)** (-Include: All columns; -Expression: lowerCase($$CURRENTCOLUMN$$); -Replace selected input columns).
2. Split the columns using multiple **Cell Splitter** (-Selection: Previous symbols/Alias symbols/UniProt ID; -delimiter: <, > & for Previous name -delimiter: <“,>; -Remove columns; - Output: new columns, Guess size and column types; *see* Fig. 2a) and **String Manipulation (Multi Column)** (-Wildcard/Regex Selection; -Pattern: <Previous name*>; -Wildcard; -Expression: removeChars($$CURRENTCOLUMN$$,"\""); -Replace selected input columns).
3. Save and export data with **CSV Writer** (-Advanced Settings: -Quotes values: Never) as HGNC_symbols_prepared.txt (*see* **Note 1**).

3.1.2 IUPHAR/BPS

1. Delete the first lines of the files from Subheading 2.2, **step 2** (ligand.csv, target_and_families.csv and interactions.csv) using text editor and load the modified files into KNIME using **CSV Readers**.
2. Join all the tables with **Joiner** (-Join columns: Ligand ID or Target ID; -Inner join; -merge join columns, -append suffix: _interaction/ _target; *see* Fig. 2b).
3. Remove columns with information you do not need with **Column Filter** and keep at least Ligand ID, Name, Type, PubChem CID, IUPAC name, Synonyms, SMILES, Target Gene Symbol, Target Species, Type_interaction, Action, Action comment, Selectivity, Affinity Units, Affinity High, Affinity Median, Affinity Low, Assay Description, PubMed ID, Type_target, Family Name, HGNC symbol.
4. Extract interactions for human targets with **Row Filter** (-Include rows by attribute value; -Column to test: Target Species; -Pattern matching; -Pattern: <Human>).

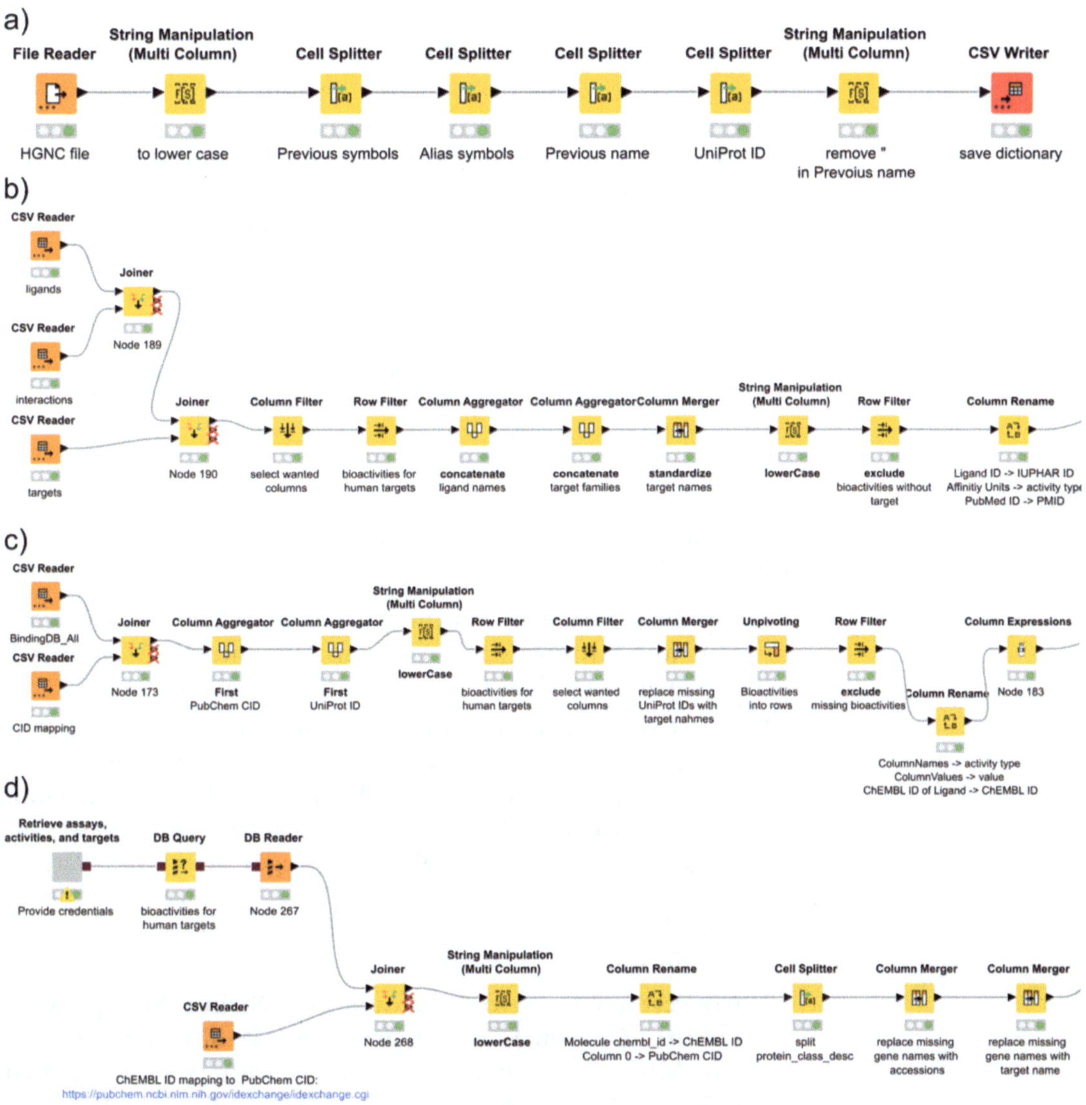

Fig. 2 Workflow steps for file preparation. (**a**) Workflow steps for preparation of the HGNC file. (**b**) Workflow steps for preparation of IUPHAR files. (**c**) Workflow steps for preparation of BindingDB files. (**d**) Workflow steps for preparation of ChEMBL database

5. Concatenate all known ligand names using **Column Aggregator** (-Aggregation column(s): Name, IUPAC name, Synonyms; -Aggregation: ligand name – Unique concatenate; -Remove aggregation columns).
6. Concatenate all known target family classifications with **Column Aggregator** (-Aggregation column(s): Type_targets, Family name; -Aggregation: target family – Unique concatenate; -Remove aggregation columns).

7. Standardize the target gene names using **Column Merger** (-Primary Column: HGNC symbol; -Secondary Column: Target Gene Symbol; -Append new column: target).
8. Convert columns to lowercase using **String Manipulation (Multi Column)** (-Exclude: SMILES, Affinity Units and columns with types other than strings; -Expression: lowerCase($$CURRENTCOLUMN$$); -Replace selected input columns) and exclude bioactivities without a target using **Row Filter** (-Exclude rows by attribute value; -Column to test: target; -only missing values match).
9. Rename columns with **Column Rename** (Ligand ID → IUPHAR ID; Affinity Units → activity type; PubMed ID → PMID).

3.1.3 BindingDB

Load the data from Subheading 2.2, **step 3** (BindingDB_All_2D_*.tsv and BindingDB_CID.txt) into KNIME with **CSV Readers.**

1. Combine bioactivity information and PubChem CID information using **Joiner** (-join columns: BindingDB MonomerID and ligand ID; -Left outer join; -merge join columns).
2. Use **Column Aggregator** (-Aggregation columns(s): PubChem CID and PubChem CID (right); -Aggregation: PubChem CID – First; -no Missing; -remove aggregation columns) to get PubChem CIDs repeat with adjusted settings to get all UniProt IDs (-Wildcard/regex Selection; -Pattern: <UniProt (SwissProt) Primary ID *>; -Wildcard; -Aggregation: UniProt ID – First; -no Missing; -remove aggregation columns; *see* Fig. 2c).
3. Convert columns to lowercase with **String Manipulation (Multi Column)** (-Exclude: Ligand SMILES column, ChEMBL ID column and columns with types other than strings; -Expression: lowerCase($$CURRENTCOLUMN$$); -Replace selected input columns).
4. Extract interactions for the human organism with **Row Filter** (-Include rows by attribute value; -Columns to test: Target Source Organism According Curator or DataSource; -Pattern: <homo sapiens>).
5. Filter columns to be included with **Column Filter** (-Include: Activity columns (Ki, IC50, EC50, Kd), ChEMBL ID, PubChem CID, Ligand name, Uniprot ID, Target Name Assigned by Curator or DataSource, PMID, BindingDB MonomerID=ligand ID).
6. Merge columns containing target names using **Column Merger** (-Primary Columns: UniProt ID; -Secondary Column: Target Name Assigned by Curator or DataSource; -Replace primary column).

7. Use **Unpivoting** (-Value Columns: -Include: Ki (nM), Kd (nM), IC50 (nM), EC50 (nM); -Retained columns: -Exclude: Ki (nM), Kd (nM), IC50 (nM), EC50 (nM)) to transpose bioactivity information from multiple columns, separated by their type, to rows.
8. Exclude rows without bioactivity value using **Row Filter** (-Exclude rows by attribute value; -Column to test: Column-Values; -only missing values match).
9. Standardize activity type, unit, and value using **Column Expressions** (1. -Replace Column; -Output column: Activity type; -Expression: replace(column("Activity type"),' (nM)' ,''); 2. -Output column: unit; Expression: 'nM'; 3. -Replace column; -Output column: value; Expression: s1 = replace(column ("value"),'>',''); s2 = replace(s1,'<','')).
10. Rename Columns with **Column Rename** (ColumnNames → Activity type; ColumnValues → value; ChEMBL ID of Ligand → ChEMBL ID).

3.1.4 ChEMBL

1. Use **PostgreSQL Connector** (-Connection Settings; -Database Dialect: PostgreSQL; -Location: -Hostname, -Port; -Database name; -Username & password; *see* Subheading 2.2, **step 4**) to connect KNIME with the database created from ChEMBL (*see* Subheading 2.2, **step 4**).
2. Use multiple **DB Table Selector** (Fig. 3) to extract the following tables: molecule_dictionary; compound_structures; compound_records; docs; activities; assays; target_dictionary; target_components; component_sequences; component_class; protein_classification; protein_family_classification.
3. Connect the sub-tables with **DB Joiner** to one global table using the ERD (*see* Subheading 2.2, **step 5**) of ChEMBL for help. The final table should contain the following columns: Molecule: chembl_id, compound_name, canonical smiles; Activities: standard_relation, standard_value, standard_units, standard_type, activity_comment, assay_description; Target: name, accession, organism, protein_class_desc, l1-l8; Docs: pubmed_id.
4. Use **DB Query** (-SQL Statement: SELECT * FROM #table# AS "table" WHERE "organism" = 'Homo sapiens') to filter activities for human targets.
5. Convert the query result table into a KNIME data table with **DB Reader**.
6. Use **CSV Reader** to load the PubChem CID mapping to ChEMBL IDs from NCBI (*see* **Note 2**) and join it with **DB Reader** using **Joiner** (-Join columns: molecule_chembl_id & Column0; -Inner join; -Merge join columns).

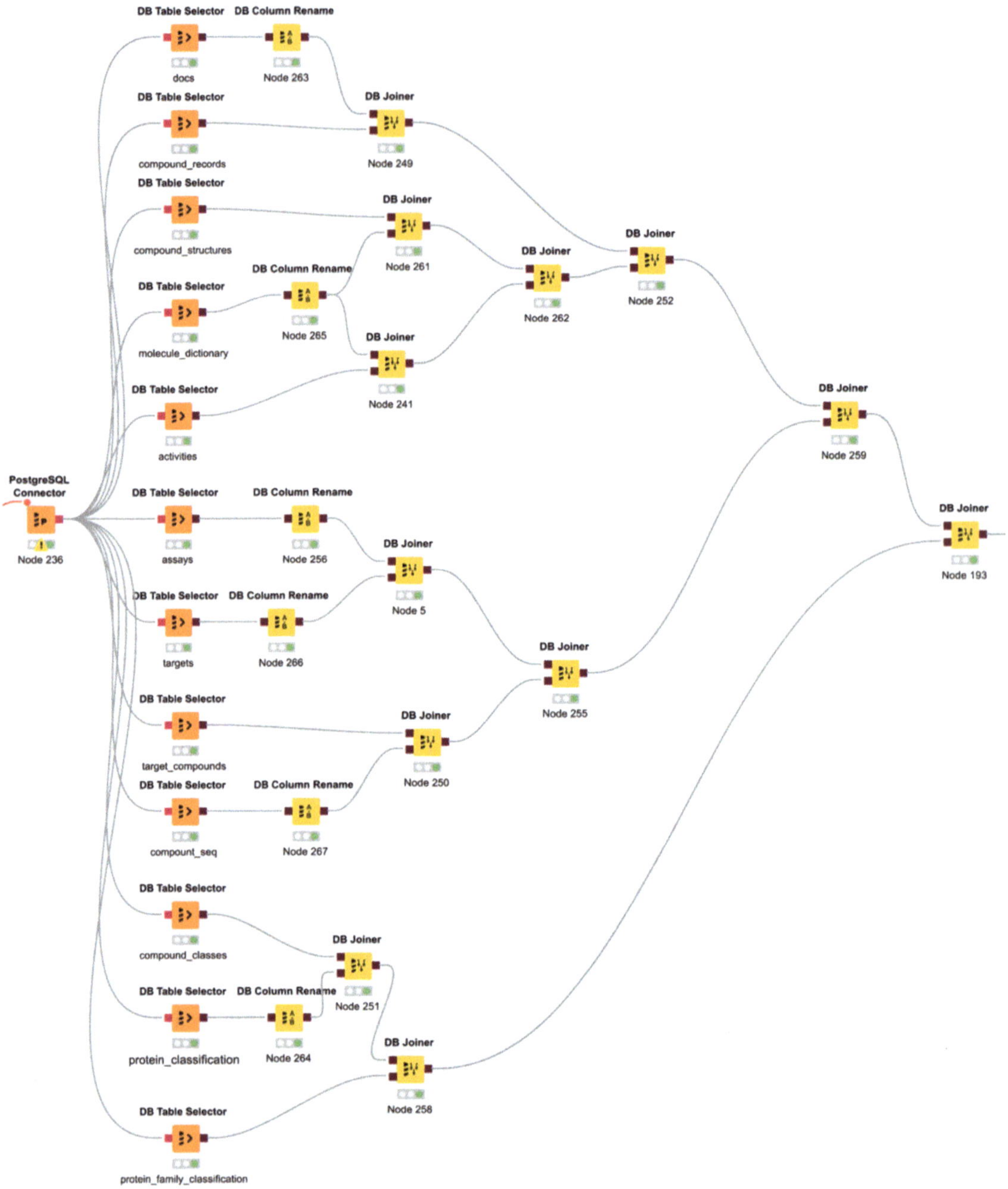

Fig. 3 Workflow steps to retrieve all information about ligands, targets and assays from the PostgreSQL database

7. Convert columns to lowercase using **String Manipulation (Multi Column)** (-Exclude: canonical_smiles, standard_type, standard_unit, molecule_chembl_id=Column0 and columns with types other than strings; -Expression: lowerCase($$CURRENTCOLUMN$$); -Replace selected input columns).

8. Standardized target names are found in the fifth substring of protein_class_desc. Thus, use **Cell Splitter** (-Select a column: protein_class_desc; -Remove input column; -delimiter: < > (2 blank spaces); -Output as new columns) to split protein_class_desc into new columns. If the protein_class_desc_Arr [5] is empty, replace it with accession (UniProt ID) or with target name using **Column Merger** (-Primary Column: protein_class_desc_Arr[5]; -Secondary Column: accession/target name; -Replace primary column) (*see* Fig. 2d).
9. Extract columns with **Column Filter** (-Exclude: protein_class_desc_Arr[1] – protein_class_desc_Arr[4], target) and use **Column Rename** (molecule_chembl_id=Column0 → ChEMBL ID; Column1 → PubChem CID; protein_class_desc_Arr[5] → target).

3.2 Data Preparation for Merging

This step is to merge compound–target data from different repositories with same activity type, assay type, and unit in only one row. Additionally, target names are standardized by using their HGNC gene symbols and bioactivity values by calculating their negative decadic logarithm (if a molar value is available). Assay types are categorized based on activity types for IUPHAR and BindingDB (pIC_{50},pEC_{50} → cell-based; pK_i,pK_d → cell-free; other → unspecified) and based on keywords for ChEMBL (e.g., binding affinity, hfret, recruitment → cell-free; reporter gene assay, luciferase → cell-based; cell proliferation, cell without any other keyword → functional). To simplify and enable an activity confidence check, we calculate the mean of bioactivity values if they deviate less than one logarithmic unit and store it along with the underlying numbers. Bioactivities with greater deviation are listed separately in multiple columns. The standardization procedure has to be performed for the datasets from all repositories. Below is an example of the data from the IUPHAR database which can be used with all other databases you wish to include.

3.2.1 Standardize Gene Names and Flag Assay Types

1. Standardize the target names using the HGNC_symbols_prepared.txt (*see* Subheading 3.1.1) and **String Replace (Dictionary)** (-Target Column: target; -Dictionary Location: <path to HGNC file>; -delimiter:<,>).
2. Extract and flag different assay conditions for IUPHAR and BindingDB using **Rule Engine** (IUPHAR/BindingDB: -Expression: $activity type$ MATCHES "pIC50" => "cell-based" \n $activity type$ MATCHES "pIC50" => "cell-based" \n $activity type$ MATCHES "pEC50" => "cell-based" \n $activity type$ MATCHES "pKi" => "cell-free" \n $activity type$ MATCHES "pKd" => "cell-free"; -Append columns: assay type) or based on keywords for ChEMBL according to **Note 3**.

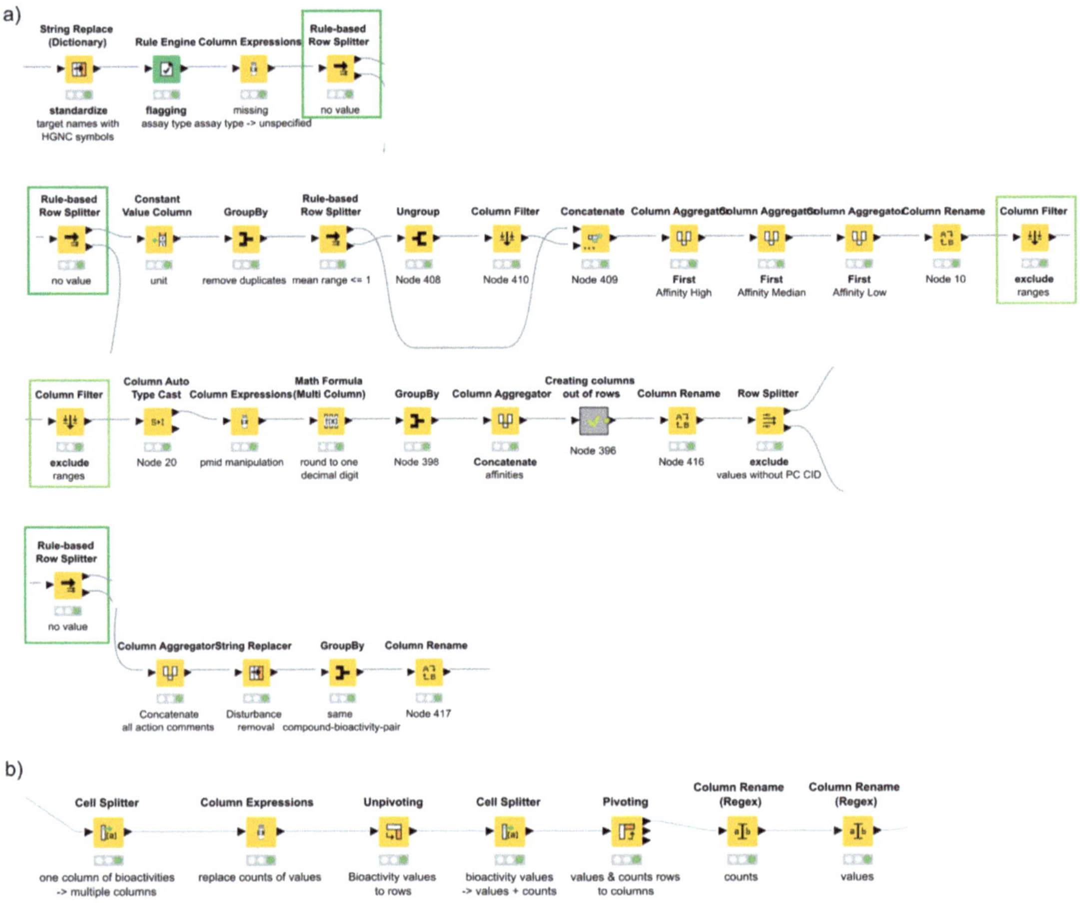

Fig. 4 Workflow steps to prepare each database for merging. (**a**) Workflow to standardize gene names and bioactivity values and to flag assay type. Colored boxes around nodes define continuing steps. (**b**) Workflow steps in metanode Creating columns out of rows

3. Assign the flag "unspecified" to all non-flagged values using **Column Expressions** (-Expression: s1 = column("assay type") \n if (isMissing(s1)) \n {'unspecified'} \n else \n {s1}; -Replace Column; -Output columns: assay type).
4. Split the table into bioactivities with and without bioactivity values. IUPHAR contains three bioactivity columns; thus, split by extracting rows where all three bioactivity columns are empty using **Rule-Based Row Splitter** (-Expression: NOT (MISSING $Affinity High$ AND MISSING $Affinity Median$ AND MISSING $Affinity Low$) => TRUE) (*see* Fig. 4a).

3.2.2 Standardize Bioactivity Values

1. The IUPHAR bioactivity values are given as negative decadic logarithm by default. Calculate the negative decadic logarithm of the bioactivity values for the other data sources according to **Note 4**. Append a column with **Constant Value Column**

(-Append: unit; -Value settings: String: neg. log.). Group the bioactivities using **GroupBy** (-Group column(s): ID, target, assay type, activity type, unit; -Aggregation settings: *see* Fig. 5a).

2. Split the table into bioactivities that are in a range of one log unit with **Rule-Based Row Splitter** (-Expression: $Range (Affinity High)$ <= 1.0 AND $Range(Affinity Low)$ <= 1.0 => TRUE \n $Range(Affinity Median)$ <= 1.0 => TRUE). For the nonmatching values (range > 1, second output table of **Rule-Based Row Filter**), use **Ungroup** (-Include: all listed columns) and **Column Filter** (-Exclude: all mean, count, range columns) to retrieve the original bioactivity values.
3. **Concatenate** the results with the first output table of the **Rule-Based Row Splitter** (*see* Fig. 4a). Replace missing values in the bioactivity columns with multiple **Column Aggregator** (e.g., -Aggregation column(s): Mean(Affinity High), List*(Affinity High); -Aggregation: Mean_I_high – First; -no Missing; -Remove aggregation columns).
4. Rename Columns to the original name (before grouping), remove the range columns using **Column Filter** (-Exclude: all range columns), and specify the column types using **Column Auto Type Cast**.
5. Change the format of the PMID column using **Column Expressions** (-Expression: s1 = replace(column ("PMID"),"[" ,"") \n s2 = replace(s1,"]" ,"") \n s3 = replace(s2,"|", "; ") \n s4 = replace(s3,",",";"); -Replace Column; -Output column: PMID).
6. Round the mean values columns using **Math Formula (Multi Column)** (-Include: Mean_I_high, Mean_I_median, Mean_I_low; -Expression: round($$CURRENT_COLUMN$$,1); -Replace selected columns). **GroupBy** (-Group column(s): IUPHAR ID, target, assay type, activity type and unit; -Manual Aggregation: *see* Fig. 5b).
7. Aggregate the bioactivity value columns using **Column Aggregator** (-Aggregation column(s): Mean_I_high, Mean_I_median, Mean_I_low; -Aggregation: Mean(-log(value)_I) – Concatenate; -Remove aggregation columns).
8. After this procedure, all the bioactivities for a compound–target pair are in one row and in one column. The mean value of matching bioactivities is only counted once and not with the correct frequency of underlying values. Thus, bring the bioactivities into one column each and replace the counts with the correct frequency of the matching bioactivities (Fig. 4b). Split the bioactivity column with the **Cell Splitter** (-column: Mean (-log(value)_I); -delimiter: <;>; -Output: as new columns;

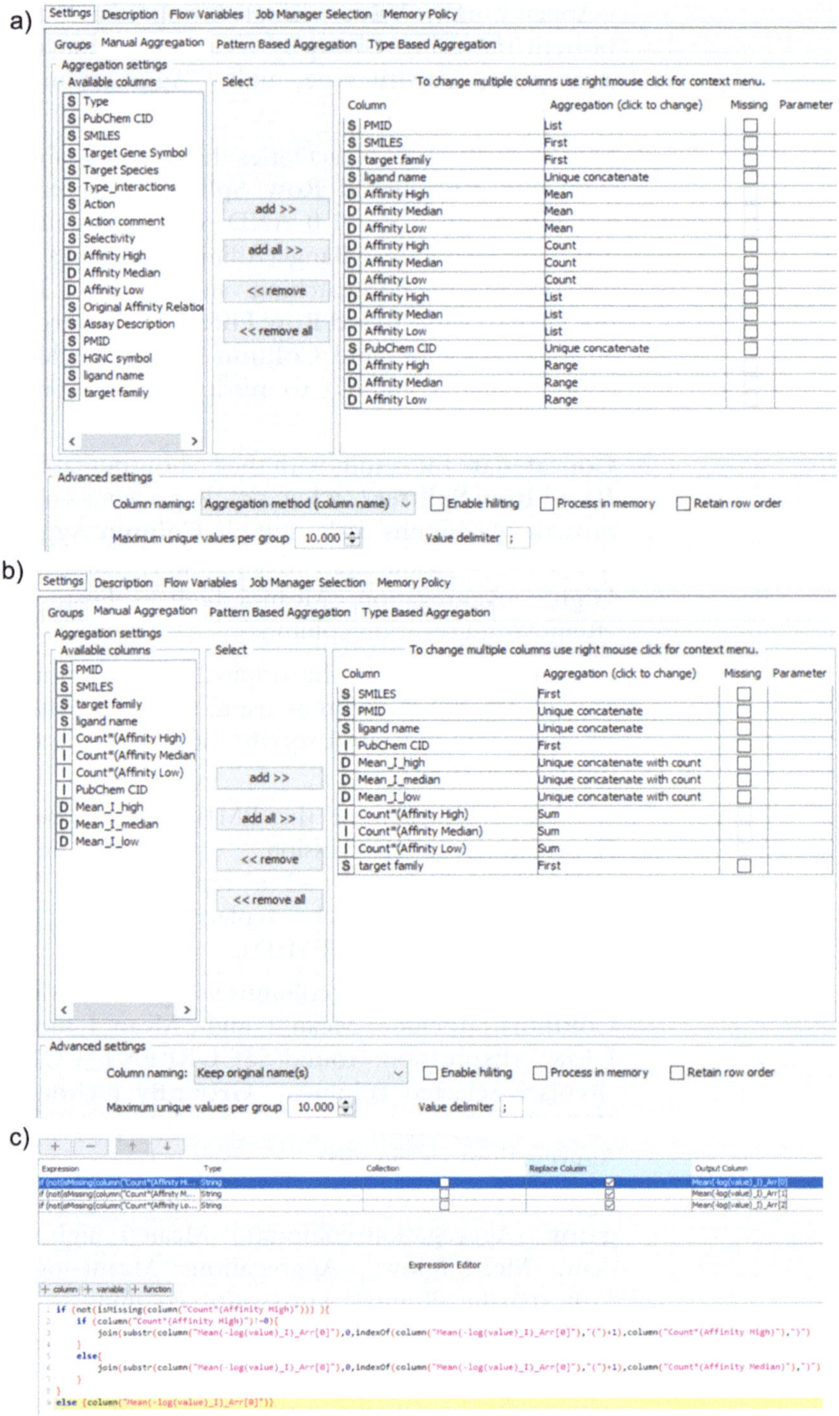

Fig. 5 Important node settings. (**a**) Settings for GroupBy *remove duplicates* to group bioactivity values and join duplicates. (**b**) Settings for GroupBy node *398*. (**c**) Example expressions for Column Expression node *replace counts of values* in metanode *Creating columns out of rows*

-Remove input columns) and replace the counts of matching values with **Column Expressions** (*see* Fig. 5c, Expressions must then be adapted to the respective database) and **Unpivoting** (Value columns: -Wildcard/Regex Selection; -Pattern: <Mean*>; Retained columns: -Wildcard/Regex Selection; -Pattern: <Mean*>; -Exclude Matching).

9. Split the new column ColumnValues with **Cell Splitter** (-Columns: ColumnValues; -delimiter: <(>; -Output as new columns; -Remove input columns) to separate values from counts and use **Pivoting** (-Group column(s): all columns, except the ColumnValues* columns, ColumnNames and Count* columns; -Pivot column(s): ColumnNames; -Aggregation: ColumnValues* columns – First).

10. Eventually, rename the bioactivity and counts columns using **Column Rename (Regex)**

    (for bioactivities: -Search String:<Mean\(log\(value\)_I\)_Arr\[(\d+)\]\+ColumnValues_Arr\[0\]>;
    -Replacement: <Mean_\[$1\]_I>;
    for counts: -Search String:<Mean\(-log\(value\)_I\)_Arr\[(\d+)\]\+ColumnValues_Arr\[1\]>;
    -Replacement: <count_value_\[$1\]_I>)

    If needed, use **Column Rename** to rename columns for distinction in the further process, e.g., *column*_I. Before merging it is critical to separate bioactivity values with known PubChem CIDs from those without PubChem CIDs, to perform a proper merging process using the **Row Splitter** (-Exclude rows by attribute value; -Column: PubChem CID; -only missing values match).

11. Some source databases (e.g., IUPHAR, ChEMBL) contain comments for records without bioactivity values. Therefore, start with the second output of **Rule-Based Row Splitter** to filter records without bioactivities out (*see* Fig. 4a). Use **Column Aggregator** (-Aggregation column(s): Action, Action comment, Selectivity; -Aggregation: Mean_I – Concatenate; -Remove aggregation columns) on to combine them to a new bioactivity column and a **String Replacer** (-Target column: Mean_I; -Wildcard pattern: -Pattern: <;> -Replacement text: -Replace all occurrences) to remove interferences.

12. Use **GroupBy** (-Group column(s): target, IUPHAR ID, assay type, activity type; -Aggregation: SMILES – First; PubChem CID – First; PMID – Unique concatenate; Ligand name – Unique concatenate; Mean_I – Unique concatenate with count) to bring all bioactivities for a compound–target pair together. If needed, use **Column Rename** to rename columns for distinction in the further process, e.g., *column*_I.

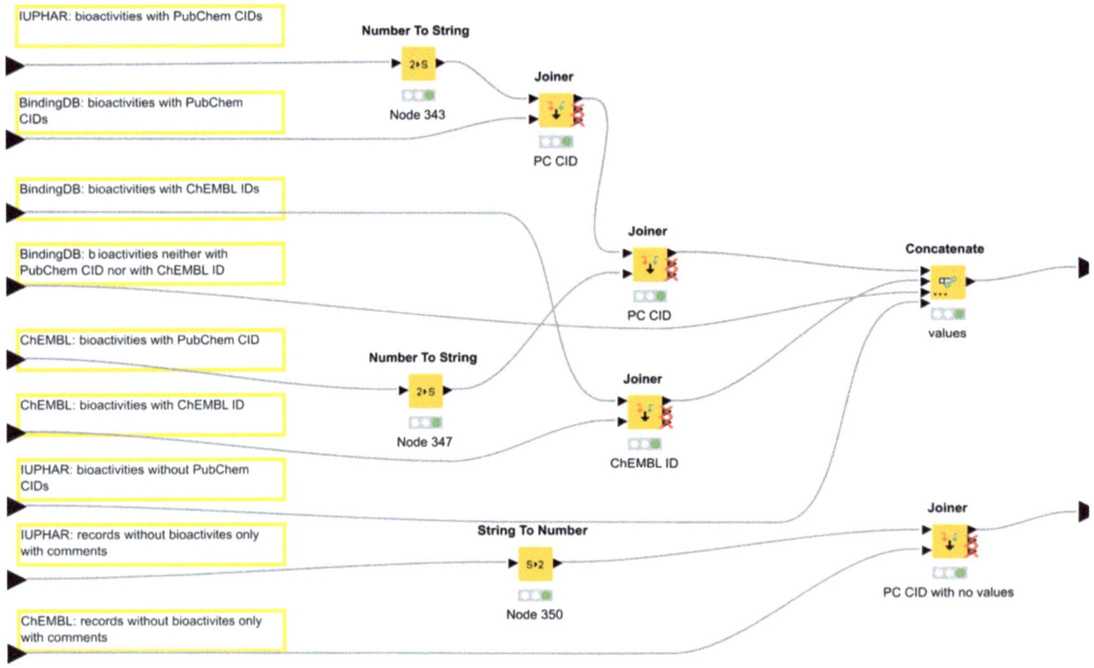

Fig. 6 Workflow steps for the merging process

3.3 Merging

1. After individual datasets preparation, combine the datasets based on the same compound, target, activity type, assay type and unit.
2. Convert the IDs (PubChem CID, BindingDB ID, IUPHAR ID) to their integer representation with **String to Number/ Number to String**.
3. Combine the databases using **Joiner** (-Join Columns: PubChem CID/ChEMBL ID, target, assay type, activity type, unit; -Full outer join; -Merge join columns) and concatenate the tables using **Concatenate**. The records with bioactivity comments can be joined by their molecule ID and target. Records with special activity types or units, for which no neg. log of the bioactivity value could be calculated, are omitted in the merging process (*see* Fig. 6).

3.4 Activity Check

After combining the different databases, we can check whether the bioactivity values from different sources match and perform an activity check for additional confidence in the bioactivity–compound–target pairs. For this purpose, we calculate the range between the minimum and maximum bioactivity value between the databases. Values within one log unit may be considered as matching, and values outside this range can be flagged (use other ranges if appropriate) (*see* Fig. 7).

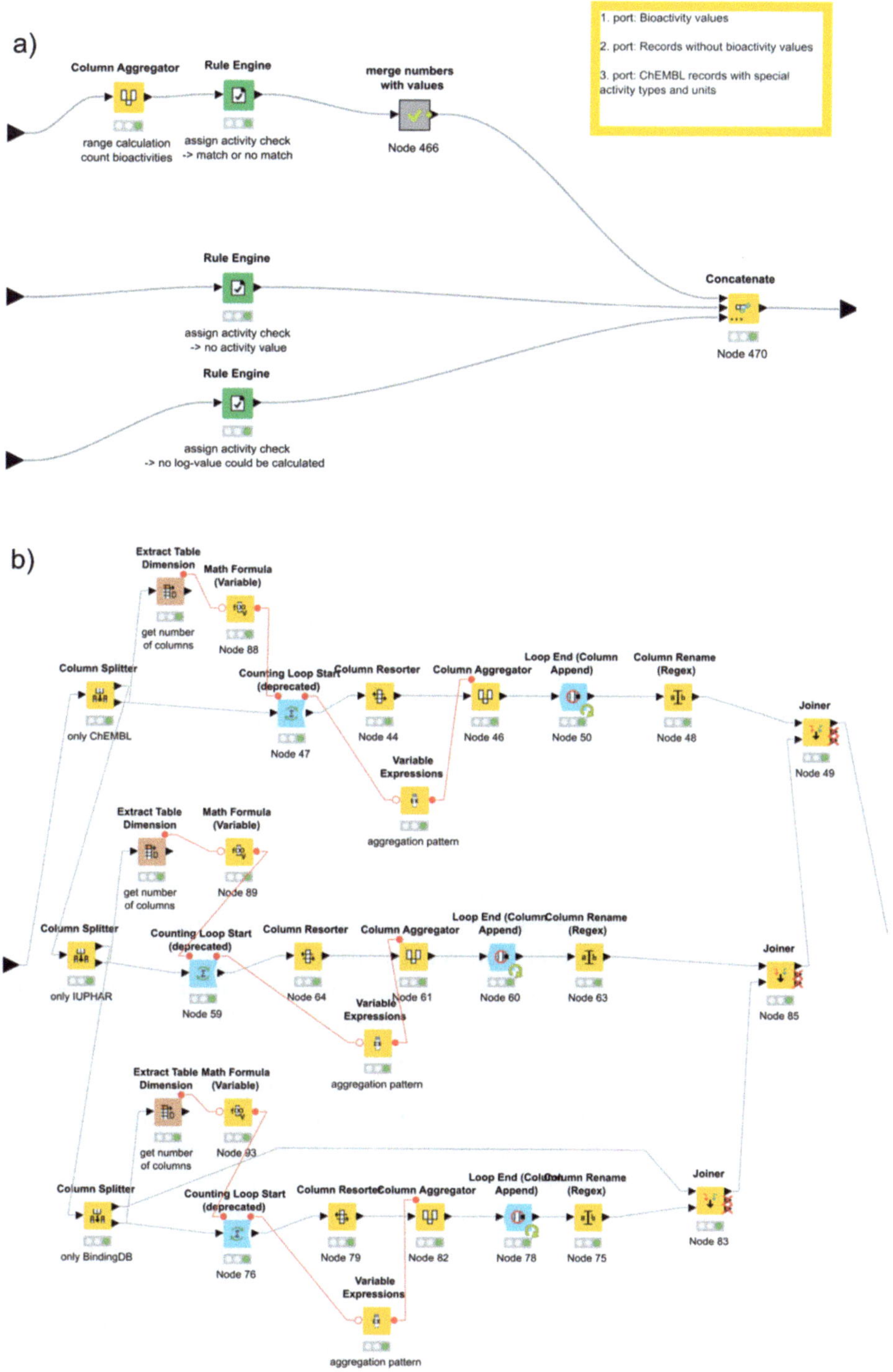

Fig. 7 Activity check. (**a**) Workflow steps for activity check. (**b**) Workflow steps to merge bioactivities with associated counts

1. Use **Column Aggregator** (Aggregation column(s): Wildcard/Regex Selection; -Pattern: <Mean*>; -Wildcard; -Aggregation: Range – Range, Count – Count; -no Missing) to calculate the range and count how many values were not missing.
2. Generate the activity flag (e.g., match, no match, only 1 data point) using **Rule Engine** (-Expression: $Count$ > 1 AND $Range$ <= 1 => " " \n $Count$ > 1 AND $Range$ > 1 => "check activity data" \n $Count$ <= 1 => "only 1 data point"; -Append column: activity check) (*see* Fig. 7a).
3. Combine the bioactivity value columns with their respective occurrence column as follows for ChEMBL as example.
4. Use **Column Splitter** (-Wildcard pattern: <*_[*]_C>) to separate ChEMBL bioactivity and count columns from the remaining table. To combine the bioactivity values with their respective count, we then group the pairs according to a pattern (*see* Fig. 7b).
5. Use **Extract Table Dimension** to retrieve the number of columns to merge and **Math Formula (Variable)** (-Expression: round($${INumber Columns}$$/2); -Replace Variable: Number Columns; -Convert to Int) to retrieve the number of final columns. Use the variable output of **Math Formula (Variable)** as variable input for **Counting Loop Start** (Flow Variables: -loops: Number Columns) and the second output of the **Column Splitter** as its table input (*see* Fig. 7b).
6. To format the resulting bioactivity columns as "bioactivity value *(counts)," resort the table using **Column Resorter** (-Actions: A-Z). Connect the Output Variable of **Counting Loop Start** with **Variable Expressions** (-Expression: join("*[",variable("currentIteration"),"]*"); -Output Variable: aggregation pattern). Connect **Column Aggregator** with **Column Resorter** and additionally connect the output of **Variable Expressions** with the Input Variable Port of **Column Aggregator** (Settings: -Wildcard/Regex Selection; -Pattern: <*[0]*>; -Wildcard; -Aggregation: Mean_C – Concatenate; -Remove aggregation columns; -Remove retained columns; -Value delimiter: <*(>; Flow Variables > aggregationColumns > name_pattern > pattern: aggregation pattern) to aggregate the desired bioactivity column with their respective count in the desired format.
7. End the Loop with **Loop End** (-Loop has same row IDs) and rename the Columns with **Column Rename (Regex)** (-Search String: <Iter #>; -Replacement: < >). Join all columns from the different databases using **Joiner** (-Join columns: Row ID; -Inner join; -Merge join columns).

8. For records without bioactivity values and with special units use **Rule Engine** (only activity comments: -Expression: TRUE => "no activity value"; special units: TRUE => "no log-value could be calculated"; -Append column: activity check).
9. Then **Concatenate** all data tables (*see* Fig. 7a).

3.5 Structure Check

Like the activity check, a structure check between the source databases may improve confidence. It can be performed by comparing the SMILES of the molecules. Structures can be flagged as matching if the SMILES match across the different databases. If SMILES from different sources do not match, their Tanimoto similarity [11] based on Morgan fingerprints [12] can point to true structural errors.

1. To perform the structure check convert the SMILES columns into SMILES representation with multiple **Molecule Type Cast** (Structure Type: Smiles) (*see* Fig. 8a) and use for each database the **RDKit Salt Stripper** (-Keep only largest fragment; -New column name: SMILES_stripped_C (e.g., for

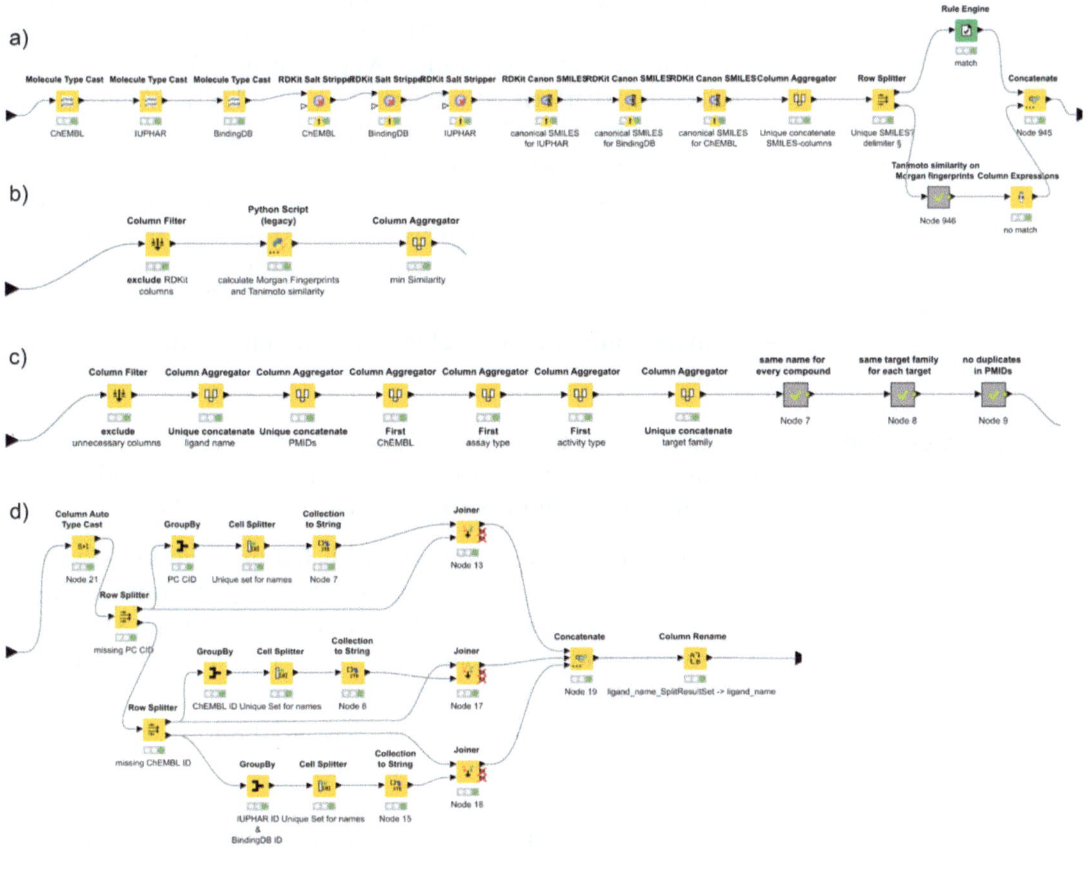

Fig. 8 Structure check and data curation. (**a**) Workflow steps for structure check. (**b**) Calculation of Morgang Fingerprints and similarity. (**c**) Workflow steps for data curation. (**d**) Workflow steps to retrieve all ligand names

ChEMBL)) to desalt the molecules. Then, canonicalize the SMILES of all databases with **RDKit Canon SMILES** (-RDKit Mol column: SMILES_stripped_C; - New column name: SMILES_C_canon (e.g., for ChEMBL)).

2. Check if the SMILES are identical between the databases using **Column Aggregator** (-Wildcard/Regex Selection; -Pattern: <*_canon>; -Wildcard; Aggregation: Unique? – Unique concatenate, Count_structures – Count; -Value delimiter:<$>). Split the rows using **Row Splitter** (-Exclude rows by attribute value; -Column: Unique?; -Pattern: <*$*>; -contains wildcards) and use **Rule Engine** (-Expression: $Count_structures $ > 1 => "match" \n $Count_structures$ = 1 => "1 structure" \n $Count_structures$ < 1 => "no structure"; -Append Column: structure check) to assign the match label to records with matching structures (first output of **Row Splitter**).
3. For the records with nonmatching structures (with delimiter, second output of **Row Splitter**), compute Morgan fingerprints (radius = 2, bits = 1024) and calculate their pairwise Tanimoto similarity using a python script (*see* **Note 7**, Fig. 8b). As python cannot handle the RDKit columns, exclude them using **Column Filter** (-Exclude: SMILES_stripped_*) before similarity calculation. Use **Column Aggregator** (-Wildcard/Regex Selection; -Pattern: <Similarity*>; -Wildcard; -Aggregation: Min_Similarity – Minimum; -Remove aggregation columns) and **Column Expressions** (-Expression: round_sim = round (column("Min_Similarity"), 2) \n join("no match (",round_sim,")"); -Output: structure check) to provide the minimum Tanimoto similarity together with structure check label *no match*.
4. **Concatenate** the two tables of matching and nonmatching structures.

3.6 Dataset Curation

After merging, activity check, and structure check, we remove unnecessary columns, retrieve all known unique ligand names, PMIDs and target families for the records (*see* Fig. 8c).

1. Remove unnecessary columns with **Column Filter** (-Exclude: RowIDs, Range, Count, Unique?, Count_structures, Min_Similarity, SMILES_stripped_* columns, original SMILES_* columns) and aggregate duplicate columns with **Column Aggregator** (*see* Fig. 8c node description).
2. Use **Column Auto Type Cast** to assign the correct data type to columns with unrecognized data type.
3. To standardize ligand names use multiple (one for each database ID (e.g., ChEMBL ID, PubChem CID, *see* Fig. 8d) **Row Splitter** (-Exclude rows by attribute value; -Column: database ID; -only missing value match) and **GroupBy** (-Group column

(s): database ID; -Aggregation: ligand_name – Unique concatenate; -Value delimiter: <;>; -Keep original name(s)).

4. Remove duplicate names using **Cell Splitter** (-Column: ligand_name; delimiter: <;>; -Output: as set; -Remove input columns) and use **Collection to String** (-Replace input columns; -seperator: \n; rest: empty) to convert the column back to string representation.
5. Join the output tables of the **Row Splitters** with the output of **Collection to String** (*see* Fig. 8d) using **Joiner** (-Join columns: database ID; -Inner join; -Merge join columns; Columns selection: -Exclude: old ligand_name) and concatenate using **Concatenate** to retrieve all ligand names.
6. Repeat the standardization procedure for target family and PMIDs.
7. Use **Column Aggregator** (-Wildcard/Regex Selection; -Pattern: <SMILES*>; -Wildcard; -Aggregation: Molecule – First; -no Missing) to get a column containing a final SMILES representation for searching molecules.
8. With **Column Resorter,** you can sort the column how do you like. The resulting database is now ready for use (*see* Subheading 3.7), e.g., for searching the known targets of a molecule or to retrieve all ligands of a target.

3.7 Application

The following procedure gives an example application of the dataset to search for potent, selective, and chemically diverse inhibitors for the glucose transporter 3 (GLUT3, SLC2A3) (*see* Fig. 9).

Fig. 9 Workflow steps for application. Dashed boxes define continuing nodes

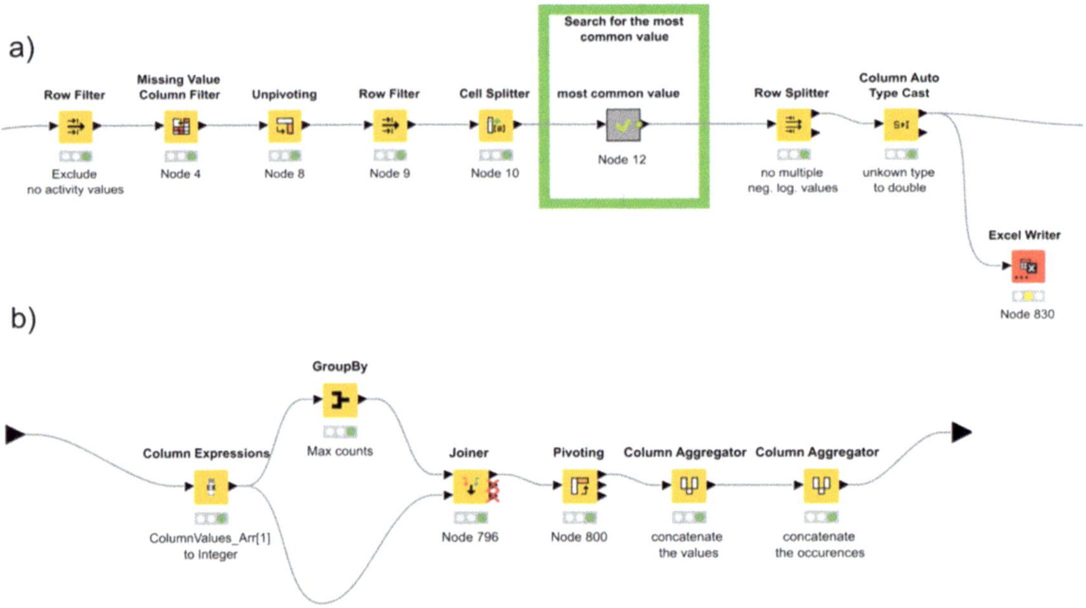

Fig. 10 Extraction of representative bioactivity values for compund-target-pair by searching for the most common value in the dataset. (**a**) Workflow steps for preparation of bioactivity values. (**b**) Workflow steps to search for the most common value

3.7.1 Search for Potent Ligands of a Target of Interest

1. Search for bioactivities for a target of interest using **Row Filter** (-Include rows by attribute value; -Column: <slc2a3> (target HGNC symbol in lower letters)). Search for K_i-, K_d-, IC_{50}- and EC_{50}-values using **Rule-Based Row Filter** (-Expression: $Activity type$ MATCHES "pKi" => TRUE \n $Activity type$ MATCHES "pKd" => TRUE \n $Activity type$ MATCHES "pIC50" => TRUE \n $Activity type$ MATCHES "pEC50" => TRUE).

2. To retrieve a representative bioactivity value, we search for the most common bioactivity as follows (as an example; *see* Fig. 10a). First, exclude records with no values using **Row Filter** (-Exclude rows by attribute; -Column: activity check; -Pattern: <No activity value>). To speed up computing time *see* **Note 5**. Then, connect the table to **Unpivoting** (Value columns: Wildcard/Regex Selection; -Pattern: <Mean_*>; -Wildcard; Retained columns: -Wildcard/Regex Selection; -Pattern: <Mean_*>; -Wildcard AND Exclude Matching). Speed up according to **Note 6**. Use the **Cell Splitter** (-Column: ColumnValues; -delimiter: <*(>; -Output: as new columns) to get bioactivity and occurrence in separate columns. Create a new **Metanode** and name it *most common value*. Open the metanode and connect the input-arrow with **Column Expression** (-Expression: removeChars(column("ColumnValues_Arr[1]"),")"); -Replace Column; -Output: ColumnValues_Arr[1]; -Type: Number (integer)). Get the

maximum occurrence of a bioactivity with **GroupBy** (-Group column(s): RowIDs; -Aggregation: ColumnValues_Arr[1] – Maximum; - Column naming: Keep original name(s)). Connect **Joiner** with **Column Expression** and **GroupBy** (-Join columns: RowIDs, ColumnValues_Arr[1]; -Inner join) (*see* Fig. 10b). After this, we pivot the table using **Pivoting** (-Group column(s): all columns except ColumnNames, ColumnValues, ColumnValues_Arr[0] and ColumnValues_Arr[1]; -Pivot column(s): ColumnNames; -Aggregation: ColumnValues – First, ColumnValues_Arr [0] – First, ColumnValues_Arr[1] – First). It is possible that multiple maximum values occur, if you want to concatenate them, use **Column Aggregator** (-Type selection: Number (double); -Aggregation: most freq. value – Concatenate (or Mean); -no Missing; -Remove aggregation columns). If also the maximum number occurs multiple times, use **Column Aggregator** (-Type selection: Number (integer); -Aggregation: occurrences – First; -no Missing; -Remove aggregation columns). If some columns have unknown datatypes, they can be corrected with **Column Auto Type Cast**.

3. If you want to find compounds with a specific potency, e.g., 1 μM or 10 μM, use **Row Filter** (-Include rows by attribute value; -Column: most freq. value; -use range; -lower bound: 6.0) (*see* Fig. 9 – first part).
4. To save all results, use **Excel Writer** or **CSV Writer** (You can review the results with the column activity check).

3.7.2 Extract Off-Targets and Find the Most Selective Compounds

1. Follow this procedure to retrieve compounds that are most selective within their target family, i.e., have as few off-targets as possible (*see* Fig. 9 – second part). Use **GroupBy** (-Group column(s): Molecule; -Aggregation: ChEMBL ID, PubChem ID, IUPHAR ID, PMIDs_Set – Unique concatenate, most freq. value – Concatenate; - no Missing; -Column naming: Keep original name(s)) to get all unique molecules to search for all annotated bioactivities and their targets.
2. To extract all annotated bioactivities for the retrieved molecules, connect **GroupBy** and the final dataset table (last node **Column Resorter** of Subheading 3.6) with **Joiner** (-Join columns: Molecule; -Inner join) (*see* Fig. 9 – second part).
3. Filter for bioactivities with K_i-, K_d-, IC_{50}- and EC_{50}-values using **Rule-Based Filter** (-Expression: $Activity type$ MATCHES "pKi" => TRUE \n $Activity type$ MATCHES "pKd" => TRUE \n $Activity type$ MATCHES "pIC50" => TRUE \n $Activity type$ MATCHES "pEC50" => TRUE).
4. Extract all bioactivities for the target family but exclude the target itself using another **Rule-Based Filter** (-Expression: $target family$ LIKE "*slc superfamily of solute carriers*"

AND NOT $Target$ MATCHES "slc2a3" => TRUE). *See* **Note 8** to find the name of the target family of interest and paste it in lowercase.

5. To retrieve off-targets of compounds of interest, repeat the steps described in Subheading 3.7.1 to obtain a representative bioactivity for the compound-off-target pair. Therefore, copy and paste all nodes from Subheading 3.7.1 searching for the most common bioactivity value and adapt as follows: in metanode *most common value*: **Pivoting** (-Group columns: additional neg. log. value) and after the metanode *most common value* insert **Column Rename** (most freq. value → most freq. value main-target; most freq. value (#1) → most freq. value off-target). Also change the **Row Splitter** (-Column: most freq. value off-target).
6. To select compounds with the lowest number of off-targets (≤ 10 μM, use other cutoff if appropriate), use **Row Filter** (-Include rows by attribute value; -Column: most freq. value off-target; -range checking; -lower bound: 5.0) and then **GroupBy** (-Group column(s): Molecule; -Aggregation: ChEMBL ID, PubChem ID, IUPHAR ID, Target – Unique concatenate; Target – Unique count; most freq. value main-target – Unique concatenate with count; most freq. value off-target – Unique concatenate with count; -no Missing). Define the number of allowed off-targets using **Row Filter** (-Include; -Column: Unique count*(Target); -range checking; -Upper bound: 2). Then, rename the columns using **Column Rename** (Unique concatenate*(Target) → off-targets; Unique count*(Target) → number of off-targets; Unique concatenate with count*(most freq. value off-target) → most freq. value off-targets; Unique concatenate with count *(most freq. value molecule) → most freq. value main-target).
7. Use the **Excel Writer** or **CSV Writer** to export and save your results.

3.7.3 Select Chemically Diverse Compounds

1. Use **Molecule Type Cast** (-Column: Molecule; -Type: Smiles) and **RDKit Find Murcko Scaffolds** (RDKit Mol column: Molecule; New column name: scaffold) to retrieve scaffolds or generic scaffolds (framework) (*see* Fig. 9 – third part).
2. Calculate Morgan Fingerprints (radius = 2, bit = 1024) on the molecules or on their scaffolds using **RDKit Fingerprint** (Fingerprint type: Morgan; RDKit Mol column: Molecule or scaffold).
3. Chemically diverse sets of molecules can then be selected manually or using the **RDKit Diversity Picker** (table 1: Molecule (Fingerprint); Number to pick: your wanted number of diverse molecules).
4. Use the **Excel Writer** or **CSV Writer** to export and save your results.

4 Notes

1. ***Preparation of HGNC File***

   The HGNC file will serve as a dictionary for the **String Replace (Dictionary)** to standardize the gene and target names provided by the various databases.

2. ***Preparation of ChEMBL***

   ChEMBL does not provide PubChem CIDs for molecules, so we use **GroupBy** on results of **DB Reader** to get ChEMBL IDs from ChEMBL without duplicates. Split the results into blocks of 500,000 ChEMBL IDs (**Row Splitter**) and save them in separate CSV files using **CSV Writer**. NCBI provides an ID exchange service to get PubChem IDs for ChEMBL IDs (https://pubchem.ncbi.nlm.nih.gov/idexchange/idexchange.cgi).

   On the website, fill in the fields as follows:

   Input ID List: Registry IDs; ChEMBL and upload the saved ChEMBL ID-file; Operator Type: Same CID; Output IDs: CIDs; Method: Two column file showing each input–output correspondence. Concatenate the files and save the resulting complete list.

3. ***Exemplary Keywords for Assay Classification by Type in ChEMBL Data Using Rule Engine***

   $assay_description$ LIKE "*binding affinity*" => "cell-free"

   $assay_description$ LIKE "*recruitment*" => "cell-free"

   $assay_description$ LIKE "*displacement*" AND NOT ($assay_description$ LIKE "*cells*" OR $assay_description$ LIKE "*luciferase*" OR $assay_description$ LIKE "*expressed in*" OR $assay_description$ LIKE "*galactosidase*" OR $assay_description$ LIKE "* luc*" OR $assay_description$ LIKE "*gal*") => "cell-free"

   $assay_description$ LIKE "*fret*" => "cell-free"

   $assay_description$ LIKE "*htrf*" => "cell-free"

   NOT ($assay_description$ LIKE "*cell*" OR $assay_description$ LIKE "*expressed in*" OR $assay_description$ LIKE "*transfection*" OR $assay_description$ LIKE "*transactivation*") => "cell-free"

   $assay_description$ LIKE "*cell*" AND $assay_description$ LIKE "*reporter*" => "cell-based"

   $assay_description$ LIKE "*cell*" AND $assay_description$ LIKE "*displacement*" => "cell-based"

   $assay_description$ LIKE "*expressed in*" AND $assay_description$ LIKE "*displacement*" => "cell-based"

   $assay_description$ LIKE "*galactosidase*" => "cell-based"

$assay_description$ LIKE "*gal*" => "cell-based"

$assay_description$ LIKE "*gal4*" => "cell-based"

$assay_description$ LIKE "*luc*" => "cell-based"

$assay_description$ LIKE "*luciferase*" => "cell-based"

$assay_description$ LIKE "*reporter gene assay*" => "cell-based"

$assay_description$ LIKE "*receptor gene assay*" => "cell-based"

$assay_description$ LIKE "*transfection*" => "cell-based"

$assay_description$ LIKE "*transactivation*" => "cell-based"

$assay_description$ LIKE "*cell proliferation*" => "functional"

$assay_description$ LIKE "*cell*" AND NOT ($assay_description$ LIKE "*reporter*" OR $assay_description$ LIKE "*transactivation*" OR $assay_description$ LIKE "*gal*" OR $assay_description$ LIKE "*gal4*" OR $assay_description$ LIKE "*galactosidase*" OR $assay_-description$ LIKE "* luc*" OR $assay_description$ LIKE "*luciferase*" OR $assay_description$ LIKE "*transfection*" OR $assay_description$ LIKE "*displacement*" OR $assay_description$ LIKE "*recruitment*" OR $assay_description$ LIKE "*fret*" OR $assay_description$ LIKE "*htrf*" OR $assay_description$ LIKE "*binding affinity*") => "functional"

4. ***Calculation of Negative Decadic Logarithm of Bioactivity Value***

Rule-Based Row Splitter (-Expression: $Activity type$ LIKE "EC*" => TRUE \n $Activity type$ LIKE "IC*" => TRUE \n $Activity type$ LIKE "Ki*" => TRUE \n $Activity type$ LIKE "Kd*" => TRUE \n $Activity type $ LIKE "Ka*" => TRUE \n $Activity type$ LIKE "ED*" => TRUE \n $Activity type$ LIKE "AC*" => TRUE), followed by **Row Splitter** for all concentration units (e.g., nM, μM, . . .). To calculate negative decadic logarithm, use **Math Formula** (e.g., for nM: -Expression: -log($value $*10^(-9)); -Append Column: -log(value)_*) for all concentrations. Change activity type using **String Manipulation** (-Expression: join("p",$Activity type$); -Replace Column: Activity type) and replace unit using **String Replacer** (-Target column: unit; -Wildcard pattern; -Pattern: concentration unit (e.g., nM), - Replace: all occurrences), followed by **Concatenate** to get the whole table back.

5. ***Acceleration of the Computing Time in Missing Value Column Filter***

Tick Wildcard/Regex Selection; Pattern: <Mean_*>; Tick Wildcard; Missing value threshold (in %): 100,00.

```python
# Copy input to output
output_table_1 = input_table_1.copy()
from rdkit import Chem, DataStructs
from rdkit.Chem import AllChem

output_table_1['Similarity_B_I'] = None
output_table_1['Similarity_B_C'] = None
output_table_1['Similarity_C_I'] = None

for index,(smi_I,smi_B) in enumerate(zip(input_table_1['SMILES_I_canon'],input_table_1['SMILES_B_canon'])):
    if smi_I is not None and smi_B is not None:
        fp1 = AllChem.GetMorganFingerprintAsBitVect(Chem.MolFromSmiles(smi_I),2,1024)
        fp2 = AllChem.GetMorganFingerprintAsBitVect(Chem.MolFromSmiles(smi_B),2,1024)
        output_table_1['Similarity_B_I'][index] = DataStructs.TanimotoSimilarity(fp1,fp2)

for index,(smi_I,smi_C) in enumerate(zip(input_table_1['SMILES_I_canon'],input_table_1['SMILES_C_canon'])):
    if smi_I is not None and smi_C is not None:
        fp1 = AllChem.GetMorganFingerprintAsBitVect(Chem.MolFromSmiles(smi_I),2,1024)
        fp2 = AllChem.GetMorganFingerprintAsBitVect(Chem.MolFromSmiles(smi_C),2,1024)
        output_table_1['Similarity_C_I'][index] = DataStructs.TanimotoSimilarity(fp1,fp2)

for index,(smi_B,smi_C) in enumerate(zip(input_table_1['SMILES_B_canon'],input_table_1['SMILES_C_canon'])):
    if smi_C is not None and smi_B is not None:
        fp1 = AllChem.GetMorganFingerprintAsBitVect(Chem.MolFromSmiles(smi_B),2,1024)
        fp2 = AllChem.GetMorganFingerprintAsBitVect(Chem.MolFromSmiles(smi_C),2,1024)
        output_table_1['Similarity_B_C'][index] = DataStructs.TanimotoSimilarity(fp1,fp2)
```

Fig. 11 Exemplary python script to compute Morgan Fingerprints and Jaccard-Tanimoto similarity in KNIME

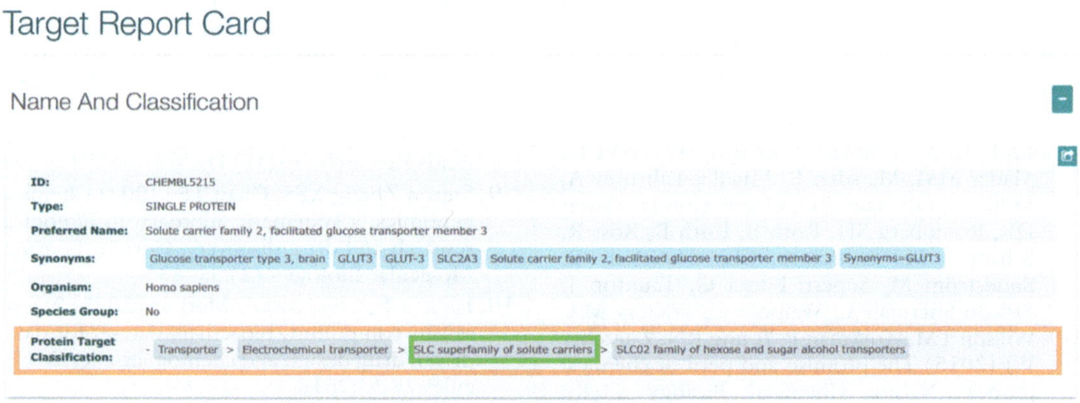

Fig. 12 Protein target classification in ChEMBL to identify target family

6. ***Acceleration of the Computing Time in Row Filter***

   Exclude rows by attribute value; Column to test = Column Values; Tick only missing values match.

7. ***Fingerprint and Similarity Calculation in Python Script***

   Morgan Fingerprints (also known as Circular or Extended Connectivity fingerprints) and the Jaccard–Tanimoto coefficient can be computed in a python script in KNIME according to Fig. 11.

8. ***Target Family:*** To find the family name of a target of interest, search for the (or a representative target) in ChEMBL and find the name under Name and Classification > Protein Target Classification (Fig. 12).

References

1. Müller S, Ackloo S, Al Chawaf A, Al-Lazikani B, Antolin A, Baell JB, Beck H, Beedie S, Betz UAK, Bezerra GA, Brennan PE, Brown D, Brown PJ, Bullock AN, Carter AJ, Chaikuad A, Chaineau M, Ciulli A, Collins I, Dreher J, Drewry D, Edfeldt K, Edwards AM, Egner U, Frye SV, Fuchs SM, Hall MD, Hartung IV, Hillisch A, Hitchcock SH, Homan E, Kannan N, Kiefer JR, Knapp S, Kostic M, Kubicek S, Leach AR, Lindemann S, Marsden BD, Matsui H, Meier JL, Merk D, Michel M, Morgan MR, Mueller-Fahrnow A, Owen DR, Perry BG, Rosenberg SH, Saikatendu KS, Schapira M, Scholten C, Sharma S, Simeonov A, Sundström M, Superti-Furga G, Todd MH, Tredup C, Vedadi M, Von Delft F, Willson TM, Winter GE, Workman P, Arrowsmith CH (2022) Target 2035 – update on the quest for a probe for every protein. RSC Med Chem 13(1):13–21
2. Arrowsmith CH, Audia JE, Austin C, Baell J, Bennett J, Blagg J, Bountra C, Brennan PE, Brown PJ, Bunnage ME, Buser-Doepner C, Campbell RM, Carter AJ, Cohen P, Copeland RA, Cravatt B, Dahlin JL, Dhanak D, Edwards AM, Frederiksen M, Frye SV, Gray N, Grimshaw CE, Hepworth D, Howe T, Huber KVM, Jin J, Knapp S, Kotz JD, Kruger RG, Lowe D, Mader MM, Marsden B, Mueller-Fahrnow A, Müller S, O'Hagan RC, Overington JP, Owen DR, Rosenberg SH, Roth B, Roth B, Ross R, Schapira M, Schreiber SL, Shoichet B, Sundström M, Superti-Furga G, Taunton J, Toledo-Sherman L, Walpole C, Walters MA, Willson TM, Workman P, Young RN, Zuercher WJ (2015) The promise and peril of chemical probes. Nature Chemical Biology 11(8): 536–541
3. Bredel M, Jacoby E (2004) Chemogenomics: an emerging strategy for rapid target and drug discovery. Nature Reviews. Genetics 5(4): 262–275
4. Jones LH, Bunnage ME (2017) Applications of chemogenomic library screening in drug discovery. Nature Reviews. Drug Discovery 16(4):285–296
5. Moffat JG, Vincent F, Lee JA, Eder J, Prunotto M (2017) Opportunities and challenges in phenotypic drug discovery: an industry perspective. Nature Reviews. Drug Discovery 16(8):531–543
6. Kim S, Chen J, Cheng T, Gindulyte A, He J, He S, Li Q, Shoemaker BA, Thiessen PA, Yu B, Zaslavsky L, Zhang J, Bolton EE (2021) PubChem in 2021: new data content and improved web interfaces. Nucleic Acids Research 49 (D1):D1388–D1395
7. Mendez D, Gaulton A, Bento AP, Chambers J, De Veij M, Félix E, Magariños MP, Mosquera JF, Mutowo P, Nowotka M, Gordillo-Marañón M, Hunter F, Junco L, Mugumbate G, Rodriguez-Lopez M, Atkinson F, Bosc N, Radoux CJ, Segura-Cabrera A, Hersey A, Leach AR (2019) ChEMBL: towards direct deposition of bioassay data. Nucleic Acids Research 47(D1): D930–D940
8. Harding SD, Armstrong JF, Faccenda E, Southan C, Alexander SPH, Davenport AP, Pawson AJ, Spedding M, Davies JA (2022) The IUPHAR/BPS guide to PHARMACOLOGY in 2022: curating pharmacology for COVID-19, malaria and antibacterials. Nucleic Acids Research 50(D1):D1282–D1294
9. Steven Zheng XFS, Chan TF (2002) Chemical genomics: a systematic approach in biological research and drug discovery. Current Issues in Molecular Biology 4(2):33–43
10. Isigkeit L, Chaikuad A, Merk D (2022) A consensus compound/bioactivity dataset for data-driven drug design and chemogenomics. Molecules 27(8):2513
11. Todeschini R, Ballabio D, Consonni V (2000) Distances and similarity measures in chemometrics and chemoinformatics. In: Encyclopedia of analytical chemistry. Wiley
12. Rogers D, Hahn M (2010) Extended-connectivity fingerprints. Journal of Chemical Information and Modeling 50:742–754

Chapter 4

Quality Control of Chemogenomic Library Using LC-MS

Václav Němec and Stefan Knapp

Abstract

In chemical biology, using compounds with incorrect identity or insufficient purity can lead to misleading biological activity data. Chemical quality control for confirmation of purity and compound identity is thus central to chemogenomics. We have established a medium-throughput LC-MS-based semi-automated quality control (QC) workflow with a minimal requirement for materials suitable for chemogenomics and other small molecule libraries. This rapid method can cover a broad chemical space of small organic compounds with diverse physicochemical properties such as polarity or lipophilicity.

Key words Liquid chromatography, Mass spectrometry, LC-MS, Chemogenomic library, Purity, Identity, Qualitative analysis, Quantitative analysis, Chemical integrity

1 Introduction

LC-MS is a standard and broadly used instrumentation for the simultaneous determination of exact molecular mass and purity of small organic molecules, biomolecules, and biological samples [1, 2]. This method can be routinely used for a qualitative and quantitative analysis of complex compound mixtures and for a quality control (QC) of substances (e.g., for pharmaceuticals, reagents for chemical synthesis, or products of chemical reactions) [3]. LC-MS is a combination of two analytical techniques, liquid chromatography (usually in high-performance (HPLC) or ultra-high-performance (UPLC) mode) and mass spectrometry, which are connected in a sequential setup [3].

High-throughput QC for the determination of identity and purity of compounds in chemogenomics libraries faces challenges in many folds. First, the limitation in the amounts usually common in large compound libraries sets a requirement for a QC method that can operate with small quantities. In addition, the chemical diversity of chemogenomics compounds needs a robust QC protocol capable of dealing with very different physicochemical parameters such as molecular weight (ca. 110–1600 Da), cLogP

Daniel Merk and Apirat Chaikuad (eds.), *Chemogenomics: Methods and Protocols*, Methods in Molecular Biology, vol. 2706, https://doi.org/10.1007/978-1-0716-3397-7_4,

(ranging between +8 and −15 cLogP), topological polar surface area (ca. 6.4–650), or molar absorption coefficient. The developed LC-MS protocol in this chapter presents a QC method that addresses both challenges: low consumption of materials (approx. 1 μL of 10 mM DMSO stock solution) and automated sample mixing pipeline allowing high-throughput measurement of diverse chemicals. In addition, the method is equipped with automated data analysis to complete a high-throughput setup.

Our LC-MS protocol presents a convenient and user-friendly compound identity and purity QC that can be applied to large compound libraries of diverse chemical and physicochemical properties. This protocol includes two methods, the "first-pass" and "second-pass" methods, which may be used sequentially. This increases the applicability of LC-MS QC analysis for chemogenomic compounds that usually have high variation in chemical and physiochemical properties. We had to compromise some aspects of qualitative and quantitative analysis as some compounds that have no or negligible absorption complicate quantitative analysis using a DAD detector. Since it is not feasible to record individual calibration curve for all samples in the high-throughput analytical format, the DAD-based quantification is not a precise determination of the sample purity. In addition, we observed the elution of some compounds from the column with no retention (retention time of ca. 1 column volume), which can affect the precise evaluation of sample purity. Another limitation stems from the incapability of verifying the correct stereo-chemical configuration or enantiomeric purity. Furthermore, identity confirmation via MSD could be difficult for compounds that are poorly ionized as cations in the positive MS mode, typically for negatively charged substances. Some of these limitations could be potentially resolved by employing additional or alternative detectors such as an evaporative light scattering detector (ELSD), which is suitable for compounds with poor or no chromophores. Nevertheless, our method has been exploited successfully for QC of a representative set of 556 chemogenomics compound candidates from the EUbOPEN consortium with high chemical diversity (clogP from −8.24 to +8.67; MW 125–1165 g/mol).

2 Materials

1. Analytical HPLC Agilent 1260 Infinity II coupled to G6125B MS detector unit with the following components: 1260 Infinity II Flexible Pump (G7104C); 1260 multisampler (G7167A; *see* **Note 1**); 1260 Infinity II Multicolumn Thermostat (G7116A); and 1260 Infinity II Diode Array Detector HS (G7117C).
2. Analytical column and column guard. Our setup uses a column (InfinityLab Poroshell 120 Bonus-RP) with the following

specifications: Bonus-RP; triple endcapped; 2.1 mm inner diameter; 100.0 mm length; 2.7 μm particle size; and 120 Å pore size. As column guard we used: UHPLC Guard 3PK, InfinityLab Poroshell 120, Bonus-RP (product number: 821725-925).

3. Acoustic liquid dispensing "Labcyte Echo 550" (Beckman).
4. The compound library (e.g., in this protocol EUbOPEN chemogenomic compounds) at 10 mM in DMSO stored in 400 μL barcoded matrix tubes.
5. 384-LDV echo plates (LP-0200).
6. 96 Round Well Microplate 4titude; 330 μL round wells, V-shaped bottom, clear PP.
7. Multichannel pipette.
8. HPLC grade acetonitrile.
9. Ultrapure water.
10. DMSO for molecular biology.
11. Formic acid (analytical grade or higher, additive to HPLC solvents).
12. Agilent software (OpenLab CDS, ChemStation Edition).
13. KNIME software (https://www.knime.com/).
14. Aluminum foil seal for 96-well plate.

3 Methods

3.1 Sample Plate Preparation

1. Transfer 1 μL of each chemogenomic compound at 10 mM in DMSO stored in 400 μL barcoded matrix tubes into each well of 96 round well microplates. Use one well in the plate for 100% DMSO for a blank run. *See* **Note 2**.
2. Seal the 96-well compound plates with aluminum foil and stored at room temperature.

3.2 LC-MS QC for Chemogenomic Compound

Due to high variation in chemical and physiochemical properties of chemogenomic compounds, we established two-tier LC-MS methods, the "first pass" and the "second pass," which should be performed in sequence. The first-pass method is suitable for data analyses of most samples owing to very general and universal conditions for sample dilution and chromatographic separation, while the "second-pass method" is optimized for compounds with higher lipophilicity and lower molar absorption coefficient. A compound with QC confirmation must pass at least one analytical method. Despite sharing the identical QC criteria, slightly different analytical conditions between both methods may lead to different QC results of a sample. Table 1 summarizes the key criteria used for purity and identity confirmation (*see* **Note 3** for application example and explanation).

Table 1
QC criteria

Main QC criteria	Relevant parameters
≥95% purity	Quantification by integration of the area under the curve for individual signals Purity monitoring primarily based on the DAD channel 245–395 nm Take other DAD channels into account during the manual analysis of results A sufficiently high signal (LC area at least 200 units) must be observed in at least one DAD channel
Correct identity	Observed and calculated exact mass is in agreement (0.1 Da tolerance) Corresponding exact mass is observed in SIM and TIC mode Elution time of the most intensive DAD signal overlaps with corresponding signals in SIM and TIC

3.2.1 The First-Pass Method

LC-MS Analytical Measurement for the First-Pass Method

1. Analyse all sample plates (prepared according to procedure in Subheading 3.1) using the First-pass method (parameters in Table 2).
2. The First-pass methods contains a multisampler sequence for sample dilution and injection that is outlined in Table 3.
3. Between individual runs, let the column equilibrate using the solvent system that is applied at the beginning the analytical run. In our case, the time equilibration time was set to 2 min.
4. Use seven channels of a diode array detector (DAD) to monitor seven distinct wavelength bands to detect the absorbance of different compounds (see Table 1). Use the DAD channel operating in the wavelength range of 245–395 nm for the quantitative evaluation (purity QC) since most (chemogenomic) compounds have absorption maxima in this range of wavelengths.
5. Use two mass selective detector (MSD) channels, one for total ion count (TIC) acquisition in the range of 100–1000 Da and one for single ion monitoring (SIM) of the mass of each individual sample. In positive mode, expect mainly $[M+H]^+$ (where M is the exact molecular mass for the most abundant isotope) of the samples in SIM and use it for qualitative analysis (identity QC).
6. Analyze LC-MS data of all samples in the 96-well plate using the method described below in "LC-MS data evaluation for the first pass method" Subheading 3.2.2 (*see* **Note 3**).

Table 2
Parameters for the first-pass method for LC-MS analysis

General parameters	Flow: 0.600 mL/min Low-pressure limit: 5.00 bar High-pressure limit: 600.00 bar Stop time (length of analytical run): 7.50 min Post time (equilibration between runs): 2.00 min
Solvents	Solvent A: water + 0.1% of formic acid Solvent B: MeCN + 0.1% of formic acid
DAD wavelengths/channels range (band mid, bandwidth)	245–395 nm (320 nm, 150 nm) 255–405 nm (330 nm, 150 nm) 220–400 nm (310 nm, 180 nm) 335–345 nm (340 nm, 10 nm) 305–315 nm (310 nm, 10 nm) 275–285 nm (280 nm, 10 nm) 249–259 nm (254 nm, 10 nm)
MSD parameters	Ionization mode: API-ES Polarity: positive Mass range: 100–1000 Da Step size: 0.10 Da Data acquisition with 2 MS channels: total ion count (TIC); single ion count ($M+H^+$)
Timetable for solvent gradient—first-pass method	0.0 min.: 95% solvent A, 5% solvent B 0.4 min.: 95% solvent A, 5% solvent B 6.3 min.: 0% solvent A, 100% solvent B 7.5 min.: 0% solvent A, 100% solvent B

Table 3
Multisampler sequence for the first-pass method

Multisampler sequence for sample dilution—first-pass method	1. Draw 8.00 µL from location "1" (vial containing solvent A) with speed of 100.0 µL/min and default offset 2. Draw 8.00 µL from location "2" (vial containing solvent B) with speed of 100.0 µL/min and default offset 3. Eject maximum volume to sample with default speed using offset −0.5 mm 4. Draw 100.00 µL from air with maximum speed 5. Eject 30.00 µL to sample with 250.0 µL/min using offset −0.9 mm 6. Eject 30.00 µL to sample with 400.0 µL/min using offset 0 mm 7. Eject 30.00 µL to sample with 250.0 µL/min using offset −0.9 mm 8. Eject maximum volume to sample with 300.0 µL/min using offset 3 mm 9. Wait 0.03 min 10. Draw 1.00 µL from sample with 70.0 µL/min using offset −0.9 mm 11. Wash needle in flushport with "S1" for 2 s 12. Inject

Table 4
QC results and categories

QC result	Conditions
Pass	Sample has ≥95% purity and correct identity
Fail	Sample does not meet at least one criterion (purity or identity)
Not detected	Sample shows very low or no signal in DAD or MSD detector, which may be justified by the physicochemical properties of the compound

LC-MS Data Evaluation for the First-Pass Method

1. Analyze the LC-MS results of all samples with the Agilent software to obtain three values: retention time, integrated area of peaks for each signal, and calculated observed mass for each DAD and MSD channel.
2. Record the results for each compound in a spreadsheet.
3. Sort the samples into three categories: pass, fail, and not detected (Table 4), based on the QC criteria in Table 1. *See also* **Note 4**.
4. Record the compounds that pass as "QC confirmation."
5. For the compounds that are assigned as "Fail" or "Not detected", re-perform LC-MS analysis using the second-pass method (Subheading 3.2.2 below).

3.2.2 The Second-Pass Method

During the method setup, we observed a few samples that did not pass the quality criteria, but after a more thorough analytical evaluation they appeared chemically intact and correct. We suspected that this could be due to specific physiochemical properties leading to either poor UV-VIS absorption leading to insufficient signal in UV-VIS detector or poor solubility during the dilution of sample prior to injection (1:1 mixture of MeCN/H_2O + 0.1% of formic acid). Therefore, we established the second-tier QC analysis, the "second-pass method," with adjustments including sample dilution, injection volume, and solvent gradient (*see* **Note 3**) for the compounds that failed or were flagged as "not detected" in the earlier QC.

LC-MS Analytical Measurement for the Second-Pass Method

1. Analyse samples that didnt pass the First-pass method (Assigned as "fail" or "not detected") using the complementary Second-pass method. The Second-pass method is similar to the first-pass method listed in Table 2 except using the different timetable for solvent gradient (Table 5) and a different multisampler sequence (Table 6).
2. The "second-pass method" includes a dilution step with only 14 μL of acetonitrile + 0.1% of formic acid in order to prevent precipitation of lipophilic compounds. In addition, we increased the injection volume by twofold (2 μL) to obtain a

Table 5
Parameters for solvent gradient specific for the second-pass method

Timetable for solvent gradient—second-pass method	0.0 min.: 95% solvent A, 5% solvent B 0.5 min.: 95% solvent A, 5% solvent B 5.2 min.: 0% solvent A, 100% solvent B 7.5 min.: 0% solvent A, 100% solvent B

Table 6
Multisampler sequence for the second-pass method

Multisampler sequence for sample dilution—second-pass method (Variations from "first-pass method" are highlighted in bold)	1. Draw *7.00 µL from location "2"* (vial containing solvent B) with speed of 100.0 µL/min and default offset 2. Draw *7.00 µL from location "4"* (vial containing solvent B) with speed of 100.0 µL/min and default offset 3. Eject maximum volume to sample with default speed using offset −0.5 mm 4. Draw 100.00 µL from air with maximum speed 5. Eject 30.00 µL to sample with 250.0 µL/min using offset −0.9 mm 6. Eject 30.00 µL to sample with 400.0 µL/min using offset 0 mm 7. Eject 30.00 µL to sample with 250.0 µL/min using offset −0.9 mm 8. Eject maximum volume to sample with 300.0 µL/min using offset 3 mm 9. Wait 0.03 min 10. *Draw 2.00 µL* from sample with 70.0 µL/min using offset −0.9 mm 11. Wash needle in flushport with "S1" for 2 s 12. Inject

stronger DAD signal for compounds with poor UV-VIS absorption. The multisampler sequence is outlined in Table 6 (*see* **Note 3**).

3. The solvent gradient was adjusted to a steeper slope of the nonpolar solvent B (MeCN + 0.1% of formic acid) as described in Table 5.
4. Use seven channels of a diode array detector (DAD) to monitor seven distinct wavelength bands to detect absorbance of different compounds (see Table 2). Use the DAD channel operating in the wavelength range of 245–395 nm for the quantitative evaluation (purity QC) since most (chemogenomic) compounds have absorption maxima in this range of wavelengths.
5. Use two mass selective detector (MSD) channels, one for total ion count (TIC) acquisition in the range of 100–1000 Da and one for single ion monitoring (SIM) of the mass of each individual sample. In positive mode, expect mainly $[M+H]^+$

(where M is the exact molecular mass for the most abundant isotope) of the samples in SIM and use it for qualitative analysis (identity QC).

6. Analyze LC-MS data using the method described in Subheading "LC-MS data evaluation for the first pass method" 3.2.1.2.

4 Notes

1. The 1260 Infinity II Multisampler has a multitude of benefits for high-throughput LC-MS QC for a large compound library. This instrument can handle vials and microplates up to 6144 samples, can inject samples at pressures up to 600 or 800 bar, and is equipped with robotic system that moves containers (e.g., microplates or vials) from the sample hotel to the central workspace for seamless automation during sample processing steps and injections. The multiwash system flushes the outside surfaces of the injection needle and uses seat backflush procedures to reduce carryover to less than 9 ppm.
2. Alternatively, prepare the 96-well compound plates from 384-well plates (384-LDV echo plates) using an acoustic liquid dispenser Labcyte Echo 550 (or a low-volume multichannel pipette).
3. The "second-pass method" includes a dilution step (multisampler sequence for sample dilution) to mix the sample with only 14 μL of acetonitrile + 0.1% of formic acid in order to prevent precipitation of lipophilic compounds. In addition, we increased the injection volume by twofold (2 μL) to obtain a stronger DAD signal for compounds with poor UV-VIS absorption. Accordingly, we also adjusted the solvent gradient to a steeper slope of the nonpolar solvent B (MeCN + 0.1% of formic acid).
4. Alternatively, use a KNIME workflow for sorting the compounds into each category.

References

1. Begou O, Gika HG, Theodoridis GA, Wilson ID (2018) Quality control and validation issues in LC-MS metabolomics. In: Theodoridis GA, Gika HG, Wilson ID (eds) Metabolic profiling, Methods in molecular biology, vol 1738. Springer, New York, pp 15–26. https://doi.org/10.1007/978-1-4939-7643-0_2
2. Korfmacher WA (2005) Foundation review: principles and applications of LC-MS in new drug discovery. Drug Discov Today 10(20): 1357–1367. https://doi.org/10.1016/S1359-6446(05)03620-2
3. Niessen WMA (2003) Progress in liquid chromatography–mass spectrometry instrumentation and its impact on high-throughput screening. J Chromatogr A 1000(1–2): 413–436. https://doi.org/10.1016/S0021-9673(03)00506-5

Chapter 5

Annotation of the Effect of Chemogenomic Compounds on Cell Health Using High-Content Microscopy in Live-Cell Mode

Amelie Tjaden, Stefan Knapp, and Susanne Müller

Abstract

The characterization of chemogenomic libraries with respect to their general effect on cellular health represents essential data for the annotation of phenotypic responses. Here, we describe a multidimensional high-content live cell assay that allows to examine cell viability in different cell lines, based on their nuclear morphology as well as modulation of small molecules of tubulin structure, mitochondrial health, and membrane integrity. The protocol monitors cells during a time course of 48 h using osteosarcoma cells, human embryonic kidney cells, and untransformed human fibroblasts as an example. The described protocol can be easily established and it can be adapted to other cell lines or other parameters important for cellular health.

Key words High-content imaging, Multiplex, Machine learning, Phenotypic screening, Cell viability

1 Introduction

In recent years, the use of chemogenomic compound libraries for the identification of biological effects associated with specific targets has gained interest in the scientific community [1–3]. The majority of compounds used in such chemogenomic libraries are well validated in terms of target-compound correlations, but information about their suitability to be used in complex biological systems is often lacking [4]. One way to prevent wrong annotation of phenotypic readouts as a result of the inhibition of specific targets or false positives in scientific research is the prior annotation of chemogenomic compounds, based on their effect on cellular health in different cellular systems. In our laboratory, we use the multiplex assay described here, as a secondary screen for compounds that show a reduced cell growth rate in a primary viability assessment, using a label-free brightfield method described in Chap. 6. There, the compounds are validated based on their

Daniel Merk and Apirat Chaikuad (eds.), *Chemogenomics: Methods and Protocols*, Methods in Molecular Biology, vol. 2706, https://doi.org/10.1007/978-1-0716-3397-7_5,

calculated growth rate (GR) [5] after 24 h. All "hit" compounds, which are classified as "cytotoxic," when containing a GR value smaller than 0, can be tested further. However, the multiplex method can be used as primary screen as well.

In the multiplex high-content assay, basic cellular functions such as cell viability, mitochondrial health, membrane integrity, and interference of compounds with the cytoskeleton are assessed [6]. It is performed in live-cell mode over a time period of 48 h, but can be easily expanded to longer time points. To avoid a too strong bias of the results, we are routinely testing three different cell lines with diverse cellular morphology: human embryonic kidney cells (HEK293T), osteosarcoma cells (U2OS), and untransformed human fibroblasts (MRC-9). Other adherent cells have been used in this assay as well.

Artificial intelligence (AI) techniques greatly improved image analysis [7] in recent years. The large amount of data that can be generated and the easier accessibility of phenotypic screening libraries have taken data evaluation to the next level. Also, morphological alterations, which are not visible by the human eye, can be detected [8].

Here, for an advanced image analysis, the data analysis is performed using a machine learning-based algorithm approach, where the cells are gated according to a "tree principle," using different characteristics as decision points. For this, we applied the CellPathfinder software from Yokogawa. The machine learning-based algorithm was trained by an experienced cell biologist, based on a set of compounds containing references for every gating step (*see* Subheading 2.2) [6]. After detection of cell body ("Cellbody") and nucleus ("Nucleus"), the cells are gated based on different features in categories connected to 1. compound properties, 2. cell properties, and 3. phenotypic properties. First, Hoechst high-intensity objects are recognized based on the Hoechst33342 channel intensity, to detect autofluorescence or precipitation of a compound. Afterwards, all remaining cells, termed "normal," are further evaluated based on their nuclear morphology and gated in either healthy, pyknosed, or fragmented. The addition of Annexin V (*see* Subheading 2.3) enables the differentiation between mitotic and apoptotic cells. For the phenotypic properties, the healthy cells are gated in three different ways: tubulin effect, increase of mitochondrial mass, and membrane permeabilized. All analysis protocols used in this protocol can be found here as follows: https://doi.org/10.5281/zenodo.6415330.

The protocol described here was validated and tested against a compound set of 230 compounds [6]. It is easy adaptable, but must then be optimized accordingly.

For more detailed information on the establishment and use of the protocol, please see Tjaden et al. [6].

2 Materials

To minimize contamination, all working steps involving live samples should be carried out under sterile conditions. All cell culture reagents, including medium, washing solution and reagents for cell passaging should be stored at 4 °C and be warmed up to 37 °C prior to use. Do not use cells with a higher passage number than 35, to ensure genomic integrity [9]. Compounds of interest, as well as reference compounds, can be prepared as aliquots in advance. They can be stored in a sealed plate or closed vial at −20 °C for a long time period. If used regularly, compounds can be stored at room temperature (RT) for 3–6 month, to avoid freeze–thaw cycles [10]. Chemical integrity, stability, and solubility of the compounds should be validated beforehand (*see* **Note 1** and Chap. 6).

2.1 Cell Culture

The described protocol is optimized and validated to test the adherent cell lines human embryonic kidney cells (HEK293T), osteosarcoma cells (U2OS), and untransformed human fibroblasts (MRC-9) at once. It can be adapted to test just one cell line or different adherent cell lines (*see* **Note 2**).

1. Cell lines: HEK293T (ATCC® CRL-1573™), U2OS (-ATCC®HTB-96™), MRC-9 fibroblasts (ATCC® CCL-2™).
2. Culture medium for HEK293T cells and U2OS cells: Dulbecco's modified Eagle's medium (DMEM) (Gibco) plus L-glutamine (High glucose) supplemented with 10% FBS (Gibco) and penicillin/streptomycin (Gibco).
3. Culture medium for MRC-9 cells: Eagle's Minimum Essential Medium (EMEM) (Gibco) supplemented with 10% FBS (Gibco) and penicillin/streptomycin (Gibco).
4. Fetal bovine serum (FBS).
5. Penicillin (100 U/mL)/streptomycin (0.1 mg/mL) (Gibco).
6. Cell washing solution: Dulbecco's phosphate-buffered saline 1X (DPBS): without calcium or magnesium (Thermo Fisher Scientific).
7. Dissociation reagent: trypsin-EDTA 1X (0.05%) (Thermo Fisher Scientific).
8. Imaging plate: 384-well cell culture microplate, PS, f-bottom, μClear® (Greiner).

2.2 Preparation of Compounds

The protocol described here is optimized to test 135 compounds at two different concentrations (1 μM and 10 μM). However, this can be adapted (*see* **Note 3**). The compounds of interest, as well as reference compounds, should be diluted in a suitable solvent, preferably DMSO, where solubility is anticipated according to the supplier. Precipitated compounds should be excluded before

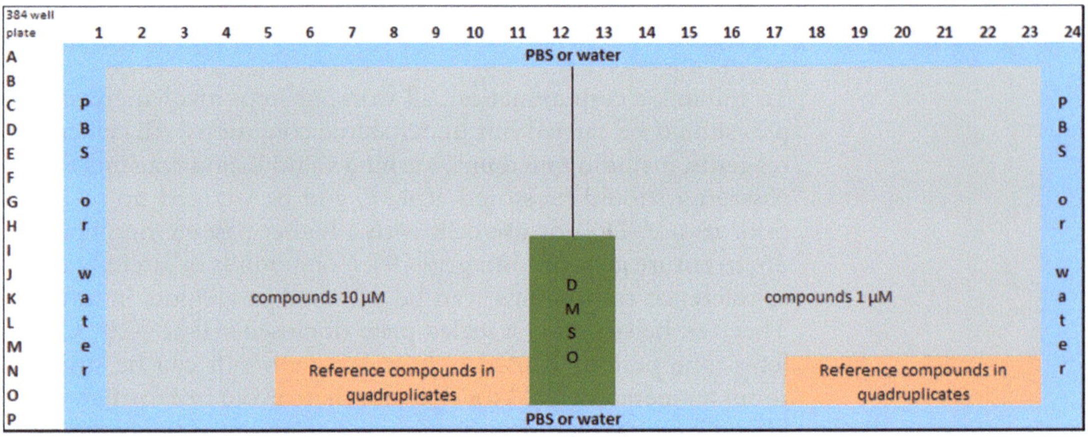

Fig. 1 Example layout of 384-well plate to test 135 compounds at two different concentrations. (This figure is adapted from Tjaden et al. [17])

testing (*see* **Note 1**). Reference compounds with known mode of actions must be added to every experiment, because they serve as training images for the subsequent analysis. As negative control, 10% of the plate should contain wells with 0.1% DMSO (*see* **Note 4**). An example layout of the plate can be found in Fig. 1. For each cell property, at least one reference compound must be tested. As part of the assay development, we tested the following reference compounds [6] and confirmed their suitability to generate a training dataset as follows:

1. Compounds are tested at 1 µM and 10 µM in technical duplicates.
2. Reference compounds should be added in quadruplicates at a final concentration of 10 µM.
   - (a) Staurosporine: apoptotic cells, pyknosed nuclei [11].
   - (b) Paclitaxel: change of tubulin structure [12].
   - (c) Milciclib: increase of mitochondrial mass [13].
   - (d) Daunorubicin: fragmented nuclei, apoptotic cells [14].
   - (e) Digitonin: permeabilized membrane [15].
   - (f) Berzosertib: Hoechst high-intensity objects [16].
3. 10% of the plate should be filled with DMSO 0.1% to serve as negative controls (*see* Fig. 1).

2.3 Cell Staining Dyes

The following dyes were validated against all three cell lines [6]. The concentrations given here should be used, as higher concentrations can affect cell viability and lower concentrations might interfere with the detection power of the analysis software (*see* **Note 5**). Using different fluorescent dyes can lead to bleed-through problems. For the here described protocol, the overlap of fluorescence emission spectra is neglectable for all dyes but the MitoTracker Red and Annexin V Alexa Fluor 680. However, this overlap does not influence the analysis, since the excitation maxima of these two dyes are well separated during the gating. *See* also [6].

1. 57 μL of 16.23 μM stock of Hoechst33342 (Thermo Fisher Scientific) (final assay concentration: 60 nM).
2. 8 μL of BioTracker™ 488 Green Microtubule Cytoskeleton Dye (EMD Millipore).
3. 12 μL of 100 μM stock of MitoTracker red (Invitrogen) (final assay concentration: 75 nM).
4. 93 μL of Annexin V Alexa Fluor 680 conjugate (Invitrogen).

2.4 Instruments for High-Content Imaging and Parameters

1. CQ1 confocal microscope (Yokogawa) for image acquisition equipped with a CO_2 incubation chamber.
2. Five channels are used for image acquisition: Ex 405 nm/Em 447/60 nm, 500 ms, 50%; Ex 561 nm/Em 617/73 nm, 100 ms, 40%; Ex 488/Em 525/50 nm, 50 ms, 40%; Ex 640 nm/Em 685/40, 50 ms, 20%; brightfield, 300 ms, 100% transmission.
3. Magnification 10×.
4. Focus and Z stacks: 7 Z stacks with a total of 55 μm spacing, focus area for a 384 well plate with a height of 14.4 ± 0.1 mm at −15.5 μm for the lowest (−20.2 μm) and highest focus of 39.7 μm (51.6 μm).
5. Cytomat2C24 incubator (Thermo Fisher Scientific).

2.5 Analysis Software

1. CellPathfinder image analysis software (Yokogawa).
2. Analysis protocols using the CellPathfinder software for all three cell lines can be found here as follows: https://doi.org/10.5281/zenodo.6415330.

3 Methods

3.1 Preparation of Cells for Live-Cell Imaging

1. Culture HEK293T and U2OS over 2 weeks in DMEM plus L-glutamine (high glucose) supplemented with 10% FBS (Gibco) and penicillin/streptomycin (Gibco).
2. Culture MRC-9 fibroblasts in EMEM plus L-glutamine supplemented with 10% FBS (Gibco) and penicillin/streptomycin (Gibco) over the same time period.
3. When cells reach a confluence of approximately 80%, count all cell lines individually using either an automated cell counter or an hemocytometer and assess the cell number and viability (viability above 95% is acceptable).

4. Dilute the cells in 16 mL media to create a cell suspension containing the desired cell number for one 384-well plate. We recommend the following cell count per well (*see* **Note 6**) as follows:
   (a) HEK293T—1500 cells/well.
   (b) U2OS—1500 cells/well.
   (c) MRC-9—1250 cells/well.
5. Add the cell staining dyes described in Subheading 2.3. to the cell suspensions.
6. Prepare a 384-well plate for every cell line (in total three plates).
7. Mix the cells with the dye solution gently and seed 50 μL per well to a 384-well plate (*see* Subheading 2.1). The outer wells should be excluded to avoid evaporation effects (*see* **Note 7**) and can be filled with 100 μL PBS or water (*see* Fig. 1).
8. Remove any air bubbles (*see* **Note 8**).
9. Leave the plate for 30 min at room temperature (RT) to allow reattachment of the cells.
10. Incubate the plate for 18–24 h at 37 °C and 5% CO_2.

3.2 Image Acquisition of Non-treated Cells

1. Switch on the CQ1 microscope system and pre-warm the incubator. Ensure that the environmental controls regulating temperature and CO_2 partial pressure are working.
2. Turn on the laser light and let it heat up for approximately 3 min.
3. The three plates are tested one after another. It is important to keep the order of the different cell lines to minimize plate variations (*see* **Note 9**).
4. Place the first plate prepared in Subheading 3.1. into the CQ1. The other plates should be kept in the incubator until they will be measured.
5. Open the CQ1 software.
6. Create a new protocol for image acquisition.
   (a) Open the "sample" section and select the used 384-well plate with a height of 14.4 ± 0.1 mm. If your plate has not already been implemented in the software, please notify the responsible person for the instrument to implement the plate (*see* **Note 10**).
   (b) Go to the "imaging" section and select the parameters described in Subheading 2.4.
   (c) Field: Set the field in the middle of the well.

   (d) Select the blue channel and click on the autofocus "AF" button to automatically search for the best cell area on 3 to 7 different wells. If using the autofocus for your experiment, it can happen that the microscope will sometimes focus on dust and will lose focus (*see* **Note 11**).
   (e) Test the set focus again on 3 to 7 different wells to check if your focus is on the correct area. Add this area as standard focus height.
7. Save your protocol.
8. Start measurement of the first plate.
9. Repeat the **steps 4** to 7 for all cell lines tested.

3.3 Image Acquisition of Treated Cells

1. Add the compounds as well as reference compounds to the first plate (*see* Subheading 2.2.).
2. Centrifuge the plate at 100× *g* for 3 min.
3. Wait 1.5 h and repeat **steps 1** and **2** for the second plate (*see* **Note 9**).
4. Wait again 1.5 h and repeat **steps 1** and **2** for the third plate.
5. Incubate the plates for 12 h after treatment at 37 °C and 5% CO_2. Other time points for image acquisition can be used. The incubation time must be adapted accordingly (*see* **Note 9**).
6. Place the first plate into the CQ1.
7. Open the measurement protocol that was used for the image acquisition of non-treated cells prepared in Subheading 3.2.
8. Measure the plate.
9. After measuring, incubate the plate again for 12 h (*see* **Note 9**).
10. Repeat **steps 6** to **8** again after 24 h and 48 h after compound treatment (*see* Subheading 3.3, **step 1**).
11. Repeat steps 6 to 10 with the other two plates, respectively.
12. A scheme of the workflow can be found in Fig. 2.

3.4 Data Analysis Using CellPathfinder Software

1. Open CellPathfinder software.
2. Convert CQ1 images as instructed by the software.
3. Open the 24 h data of one cell line first to create an analysis protocol for this cell line (*see* **Note 12**) or use the provided protocols (*see* Subheading 2.4, **step 2**). The following steps should be performed accordingly with the other cell lines.
4. The following channels are used for the analysis as follows:
   (a) CH1: Hoechst33342 (DNA detection): Ex 405 nm/Em 447/60 nm, MaxIP.
   (b) CH2: BioTracker™ 488 Green Microtubule Cytoskeleton Dye (tubulin stain): Ex 488/Em 525/50 nm, MaxIP.

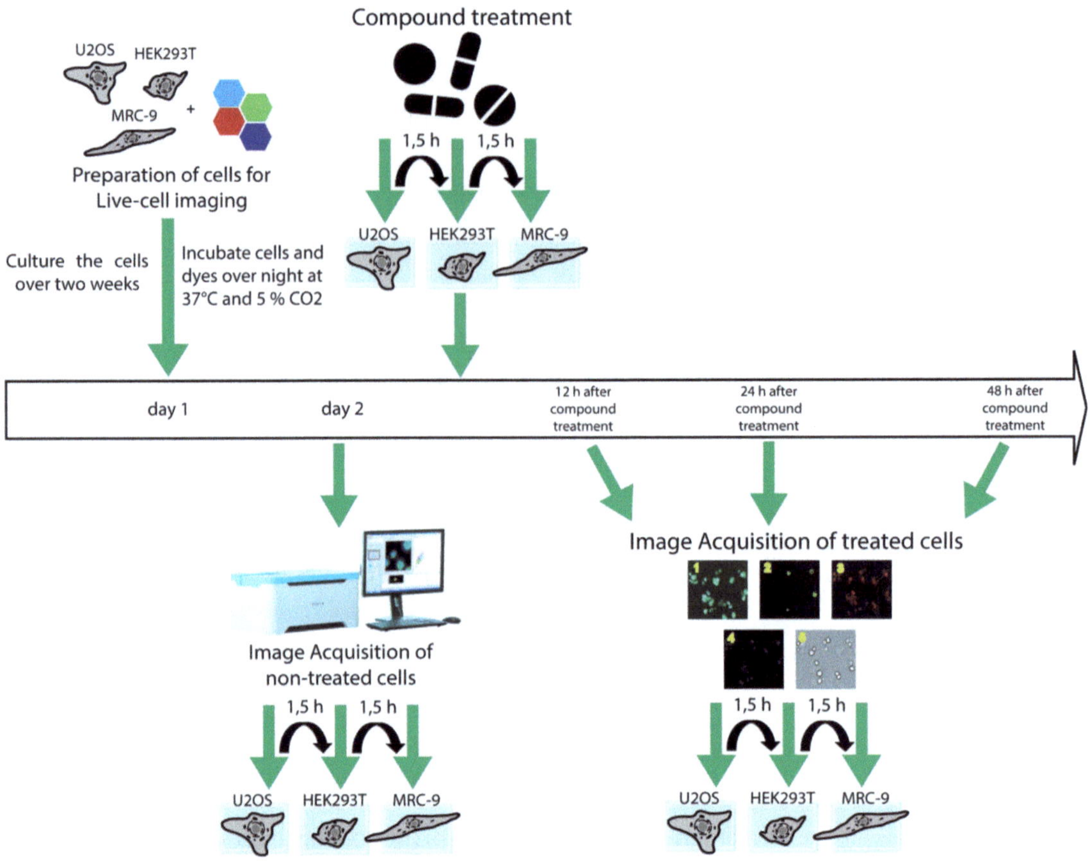

Fig. 2 Scheme of workflow for three cell lines

(c) CH3: MitoTracker red (mitochondrial mass detection): Ex 561 nm/Em 617/73 nm, MaxIP.

(d) CH4: Annexin V (apoptosis marker): Ex 640 nm/Em 685/40, MaxIP.

(e) CH5: brightfield, DCP mode "fluor-" type.

5. For object detection, create the two objects "Cellbody" and "Nucleus" (*see* Fig. 3).

   (a) "Nucleus": Ex 405 nm/Em 447/60 nm, Finder: Nuclear, Recognition: Advance.

   (b) "Cellbody": brightfield, Finder: Cell, Recognition: Advance.

6. Define the object "Nucleus" to be included in the "Cellbody" (*see* Fig. 3).

7. After object detection, the gating is performed using the machine learning (ML) based function implemented in the software.

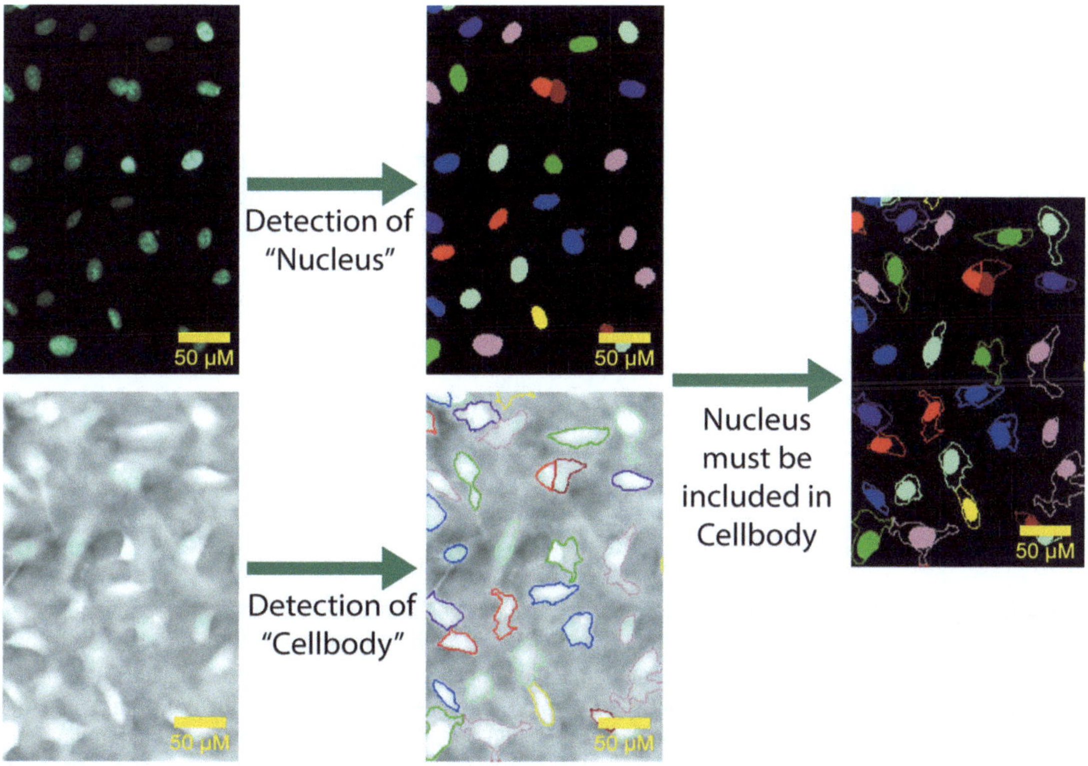

Fig. 3 Object detection. The two objects "Cellbody" and "Nucleus" should be detected by the software algorithm. The Nucleus should be defined as included in the Cellbody

(a) Use DMSO 0.1% wells as negative controls (*see* **Note 4**).

(b) Use the reference compounds (*see* Subheading 2.2) to train the ML algorithm (*see* **Note 13**).

(c) To validate your algorithm, check other wells containing the reference compounds (*see* **Note 13**).

8. The gating hierarchy can be found in Fig. 4.
9. All features used for training the machine learning algorithm are shown in Table 1.
10. Select 10 to 15 cells for every gating step. A detailed explanation of how to train the machine learning-based algorithm can be found in the CellPathfinder user information as well as in the recently published protocol by Tjaden et al. [17].
11. Save the analysis protocol.
12. Analyze the data from all time points using the same analysis.
13. Repeat the **steps 3** to **12** for the other cell lines.

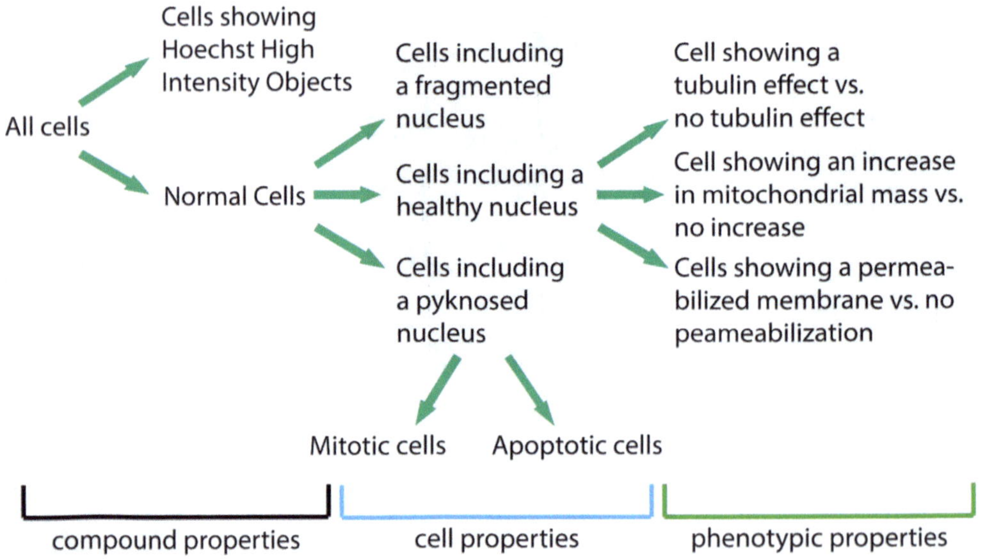

Fig. 4 Gating hierarchy. The gating is performed employing a tree principle. First, the compounds properties (Hoechst high-intensity objects or normal cells) are defined. Afterward, the cells are gated based on their nuclear morphology in healthy, pyknosed, or fragmented. Cells containing a pyknosed nucleus are further gated in apoptotic or mitotic cells. Lastly, phenotypic properties are considered to divide cells into the categories tubulin effect/no tubulin effect, mitochondrial mass increased/no increase, and membrane permeabilized/no permeabilization

3.5 Data Evaluation

1. After analyzing all cell lines for all time points, results are evaluated.
2. You can save the images under the "view" tab.
3. Growth rate can be calculated as shown in Chap. 6.
4. For biological data, at least two biological replicates should be performed.
5. The cell count of all gating steps can be used to calculate the ratios of the different properties (*see* **Note 14**) (*see* Fig. 5).
6. All data should always be compared with the screened positive and negative controls.
7. For a basic annotation, we defined the threshold of 50% to define a compound as "hit" compound. Compounds that are "hits" in more than one cell line should be evaluated further. The phenotypic evaluation can be optimized further if needed for more in-depth annotation regarding mitochondrial and tubulin features.

Table 1
Gating features used for machine learning algorithm

Gating type	Cell region	Features	Gating type	Cell region	Features
Hoechst high-intensity objects vs. normal cells	Cellbody	Mean intensity CH1	**Healthy, pyknosed, or fragmented nuclei**	Nucleus	Total intensity CH1
	Cellbody	Min intensity CH1		Nucleus	Total hole CH1
	Cellbody	Max intensity CH1		Nucleus	Total valley CH1
	Nucleus	Max intensity CH1		Nucleus	Total edge CH1
	Nucleus	Nuc_cell_area		Nucleus	Total saddle CH1
	Nucleus	Nuc area/cellbody area		Nucleus	Mean hole CH1
Mitotic vs. apoptotic cells	Cellbody	Total intensity CH4		Nucleus	Mean ridge CH1
	Cellbody	Max intensity CH4		Nucleus	Mean valley CH1
	Cellbody	Mean peak CH4		Nucleus	Mean edge CH1
	Cellbody	Mean hole CH4		Nucleus	Mean saddle CH1
	Cellbody	Mean ridge CH4	**Tubulin effect vs. no tubulin effect**	Cellbody	Total intensity CH2
	Cellbody	Mean valley CH4		Cellbody	Mean intensity CH2
	Cellbody	Mean saddle CH4		Cellbody	Max intensity CH2
	Nucleus	Total hole CH1		Cellbody	Mean peak CH2
	Nucleus	Mean peak CH1		Cellbody	Mean ridge CH2
	Nucleus	Mean hole CH1		Cellbody	Mean edge CH2
	Nucleus	Mean ridge CH1	**Membrane permeabilized vs. normal**	Cellbody	Total intensity CH2
	Nucleus	Mean valley CH1		Cellbody	Total intensity CH4
Mitochondrial mass increased vs. normal	Cellbody	Total intensity CH3		Cellbody	Mean intensity CH2
	Cellbody	Mean intensity CH3		Cellbody	Mean intensity CH3
	Cellbody	Max intensity CH3		Cellbody	Max intensity CH2
	Cellbody	Total peak CH3		Cellbody	Total intensity CH2
	Cellbody	Mean saddle CH3		Cellbody	Mean peak CH2
	Cellbody	Mean peak CH3		Cellbody	Mean peak CH3
	Cellbody	Mean hole CH3		Cellbody	Mean peak CH5
	Cellbody	Mean ridge CH3		Cellbody	Mean edge CH2
	Cellbody	Mean valley CH3		Cellbody	Mean hole CH2
				Cellbody	Mean hole CH3
				Cellbody	Mean ridge CH2
				Cellbody	Mean ridge CH3
				Cellbody	Mean ridge CH5
				Cellbody	Mean valley CH2
				Cellbody	Mean valley CH3
				Nucleus	Compactness

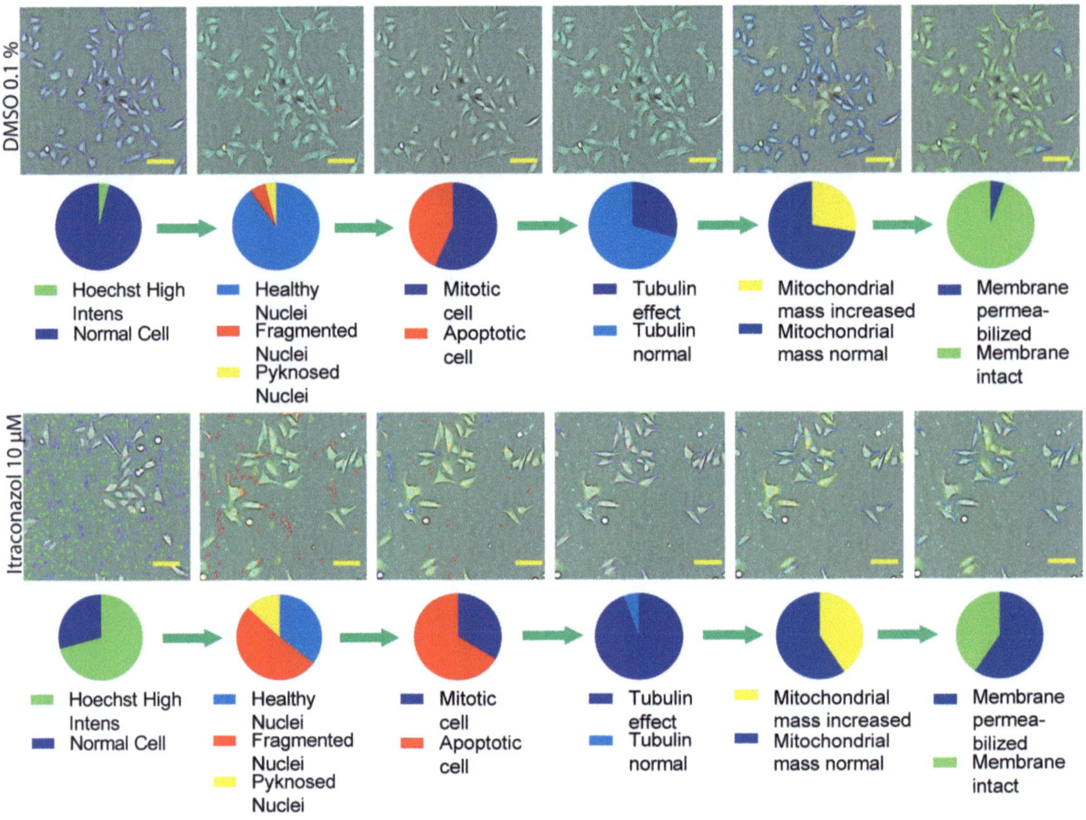

Fig. 5 Processed image of stained (blue: DNA/nuclei, green: microtubule, red: mitochondria content, magenta: Annexin V apoptosis marker) U2OS cells after 24 h of exposure to Itraconazol (10 μM) and 0.1% DMSO. U2OS cells are gated into different categories using the CellPathfinder analysis. Pie charts show the ratios of the different gating steps. (This figure is adapted from Tjaden et al. [17])

4 Notes

1. All compounds should be checked by eye if precipitation happened during storing. Precipitated compounds look cloudy in the dilution or on the plate. If precipitation happened, we recommend to freshly dissolve the compound or use a different solvent, e.g., water. A more professional check can be performed as described in Chaps 4 and 6.
2. If other cell lines than the once described here are used, the protocol must be adapted accordingly. The dye concentration, incubation time, and cell seeding density should be tested using a reference compound set. For more information on how to perform this, please see Tjaden et al. [6].
3. Other compound concentrations can be tested. If more than two are tested, the plate layout (*see* Fig. 1) must be adapted accordingly.

4. The reference compounds are used to train the machine learning-based algorithm. They should be added in quadruplicates to validate the analysis in different reference compound wells. Other compounds can also be used, when they have a similar mode of action. The machine learning-based algorithm has to be trained accordingly. The negative control is important to normalize your results and to evaluate the ratios calculated based on the gating steps in comparison to the controls.
5. All dyes used in the protocol described here were tested at different concentrations over time in all three cell lines. The concentration used was validated as the best concentration for a robust fluorescent readout without having an impact on cell viability. If other dyes or concentrations are used, the protocol must be adapted accordingly.
6. The cell count per well was tested in a pre-experiment to detect the best cell seeding density for testing the cells over 48 h. For more information about how to select the best cell seeding concentration, *see* Tjaden et al. [17].
7. Evaporation of the media and "edge-effects" can be minimized when filling the outer wells with a buffer of choice, e.g., DPBS or water in a volume of 100 μL per well.
8. Air bubbles can interfere with cell growth or later image acquisition. They should be removed using a "debubbler" or something similar. The debubbler can be a nozzle inserted into an empty bottle containing, e.g., ethanol. The nozzle should not touch the liquid, so that air can be expelled to remove any bubbles.
9. For image acquisition of one 384-well plate, the CQ1 confocal microscope approximately needs 1.5 h. This time difference should be kept in mind when pipetting the compounds. The plate order when testing different plates should always be the same. The time points of image acquisition can be changed if desired. For three plates the time points in between measurements must be more than 4.5 h.
10. If another plate is used than implemented in the CQ1, focus might be lost during the measurement. We recommend to add the plate information to the CQ1 software before measurement. This can be done by the Yokogawa support team.
11. When using autofocus for the whole measurement, the CQ1 sometimes focuses on dust on the lid instead of cells and loses the best focus area. We recommend to use the autofocus to search for the best cell area and enter this focus area as a standard for the whole plate.

12. The cell lines differ in their morphology and fluorescence intensity levels. Therefore, individual analysis protocols should be created for each cell line.
13. For the machine learning based algorithm, approximately 10 to 15 example cells of the reference compounds should be selected. To validate the analysis, other wells of the same reference compounds should be used.
14. Keeping the tree principle in mind, cell count ratios of the different gating steps can be calculated by dividing the cell count of a gating step by the cell count of next higher level. Cell count ratios can then be compared with control wells.

Acknowledgement

The authors are grateful for support by the Structural Genomics Consortium (SGC), a registered charity (No: 1097737) that receives funds from Bayer AG, Boehringer Ingelheim, Bristol Myers Squibb, Genentech, Genome Canada through Ontario Genomics Institute, Janssen, Merck KGaA, Pfizer, and Takeda and by the German Cancer Research Center DKTK and the Frankfurt Cancer Institute (FCI). This project has received funding from the Innovative Medicines Initiative 2 Joint Undertaking (JU) under grant agreement No 875510. The JU receives support from the European Union's Horizon 2020 research and innovation program and EFPIA and Ontario Institute for Cancer Research, Royal Institution for the Advancement of Learning McGill University, Kungliga Tekniska Hoegskolan, Diamond Light Source Limited. Disclaimer: This communication reflects the views of the authors, and the JU is not liable for any use that may be made of the information contained herein. A.T. is supported by the Deutsche Forschungsgemeinschaft (DFG, German Research Foundation) – grant number 259130777 (SFB1177). The CQ1 microscope was funded by FUGG (INST 161/920-1 FUGG). We thank Robert Giessmann for his help optimizing the assay and Martin Schroeder for the inspiration to perform the assay.

References

1. Bredel M, Jacoby E (2004) Chemogenomics: an emerging strategy for rapid target and drug discovery. Nat Rev Genet 5(4):262–275
2. Caron PR et al (2001) Chemogenomic approaches to drug discovery. Curr Opin Chem Biol 5(4):464–470
3. Wells CI et al (2021) The kinase Chemogenomic set (KCGS): an Open Science resource for kinase vulnerability identification. Int J Mol Sci 22(2):566
4. Gerry CJ et al (2016) Real-time biological annotation of synthetic compounds. J Am Chem Soc 138(28):8920–8927
5. Hafner M et al (2016) Growth rate inhibition metrics correct for confounders in measuring sensitivity to cancer drugs. Nat Methods 13(6): 521–527

6. Tjaden A et al (2022) Image-based annotation of chemogenomic libraries for phenotypic screening. Molecules 27(4):1439
7. Boyd J, Fennell M, Carpenter A (2020) Harnessing the power of microscopy images to accelerate drug discovery: what are the possibilities? Expert Opin Drug Discovery 15(6): 639–642
8. Ziegler S, Sievers S, Waldmann H (2021) Morphological profiling of small molecules. Cell Chem Bio 28(3):300–319
9. Hughes P et al (2007) The costs of using unauthenticated, over-passaged cell lines: how much more data do we need? BioTechniques 43(5):575–586
10. Kozikowski BA et al (2003) The effect of room-temperature storage on the stability of compounds in DMSO. J Biomol Screen 8(2): 205–209
11. Bruno S et al (1992) Different effects of staurosporine, an inhibitor of protein kinases, on the cell cycle and chromatin structure of normal and leukemic lymphocytes. Cancer Res 52(2): 470–473
12. Wang TH, Wang HS, Soong YK (2000) Paclitaxel-induced cell death: where the cell cycle and apoptosis come together. Cancer 88(11):2619–2628
13. Sanchez-Martinez C et al (2015) Cyclin dependent kinase (CDK) inhibitors as anticancer drugs. Bioorg Med Chem Lett 25(17): 3420–3435
14. Al-Aamri HM et al (2019) Time dependent response of daunorubicin on cytotoxicity, cell cycle and DNA repair in acute lymphoblastic leukaemia. BMC Cancer 19(1):179
15. Styrt B, Johnson PC, Klempner MS (1985) Differential lysis of plasma membranes and granules of human neutrophils by digitonin. Tissue Cell 17(6):793–800
16. Fokas E et al (2012) Targeting ATR in vivo using the novel inhibitor VE-822 results in selective sensitization of pancreatic tumors to radiation. Cell Death Dis 3(12):e441–e441
17. Tjaden A et al (2022) High-content live-cell multiplex screen for chemogenomic compound annotation based on nuclear morphology. STAR Protocols 3(4):101791

Chapter 6

Characterization of Cellular Viability Using Label-Free Brightfield Live-Cell Imaging

Lewis Elson, Amelie Tjaden, Stefan Knapp, and Susanne Müller

Abstract

In recent years, the assembly and annotation of chemogenomic libraries have gained interest by the phenotypic screening community. Apart from basic annotations of the compound potency and selectivity, these compound libraries benefit in particular from annotation regarding the effect of the inhibitors on cellular viability to distinguish between on-target effects of a compound and unspecific cytotoxicity. Here, we provide a protocol to determine viability as a first determinant in compound quality control, using the Incucyte live-cell imaging system. The compounds are classified according to their calculated growth rate to determine a cytotoxic, cytostatic, or healthy outcome. All compounds affecting the growth rate can be further evaluated regarding their specific effects on cell health in a high-content live-cell multiplex assay, described in Chapter 5.

Key words Viability assay, Phenotypic screening, Cytotoxicity, Growth rate

1 Introduction

The use of well-characterized screening libraries has garnered a rise in popularity over recent years. These libraries ideally consist of compounds with high target specificity, so-called chemical probes, or compounds with a narrower (usually restricted to a few members of a family) but not exclusive selectivity profile, so-called chemogenomic compounds [1]. The latter has seen increased interest due to the current lack of readily available chemical probes [2]. With the resurgence of phenotypic screening over traditional target-based drug discovery efforts, as it offers the advantage of not requiring the full understanding of a specific mode of action and being more physiologically relevant [3], these libraries can offset the lack of mechanistic insight that can complicate hit validation, often associated with such approaches [1]. While these libraries offer a way to

Daniel Merk and Apirat Chaikuad (eds.), *Chemogenomics: Methods and Protocols*, Methods in Molecular Biology, vol. 2706, https://doi.org/10.1007/978-1-0716-3397-7_6,

validate targets associated with a specific phenotype, it is important to note that non-specific compound toxicity will impact phenotypic readouts and thus the ability to interpret the data and associated phenotypes with the relevant molecular targets [1]. With this in mind, we propose that each compound should be characterized based on its effect on cells, in particular its impact on cell viability.

A frequently employed way to assess and profile small molecules based on their effects on cellular viability are colorimetric or fluorescent-based assay systems such as alamarBlue™ or the tetrazolium dye 3-(4,5-dimethylthiazol-2-yl)-2,5-diphenyltetrazolium bromide (MTT) [4]. In contrast to these plate reader-based approaches, live-cell imaging systems can assess cell viability without the use of additional reagents that might in turn affect cell growth. Here, cellular confluency is measured over a defined period of time before and after compound treatment. For adherent cells, confluency refers to the percentage of the culture vessel's surface area that is covered by a layer of cells. Suspension cell confluency can be determined in a similar manner (how crowded a defined area is), but is usually assessed via medium turbidity and colorimetric assessment.

The confluency parameter is a simple and cost-effective way to quantify a compound's effect on the cell. To validate the confluency independent of both cell division rates and cell seeding densities while also including time-dependent changes of compound effects, test compounds can be evaluated based on their growth rate. The growth rate is a normalized parameter (against the negative control) that provides a quantitative measure by incorporating initial cell confluency allowing for the characterization of either cytotoxic or cytostatic effects [5].

Calculated growth rate values lie in a range between "−1" and "1." Between "0" and "1," a compound shows partial inhibition of the cell growth, "0" defines a cytostatic effect, while values between "0" and "−1" define a cytotoxic effect (Fig. 1a). A value of "1" matches the value for the negative control and thus denotes no effect. Compounds that are classified as cytotoxic have growth rates lower than 1 (Fig. 1b) or appear to show a distinct morphological effect (Fig. 2) are considered as "hits" and can be further evaluated in other assays, if appropriate, such as a high-content live-cell multiplex assay (Chapter 5) which aims to elucidate the particular effect of the compound within the cell.

Several live-cell imaging systems are available, such as the CELLCYTE X™ Live-Cell Imager (CYTENA), which can be used for the assay; however, here, we describe the workflow for a one-shot assay capable of characterizing general cellular viability using the Incucyte Live Cell imaging system. With this protocol, screening of 308 small molecules (including controls) on a 384-well microplate format is possible. The main procedures

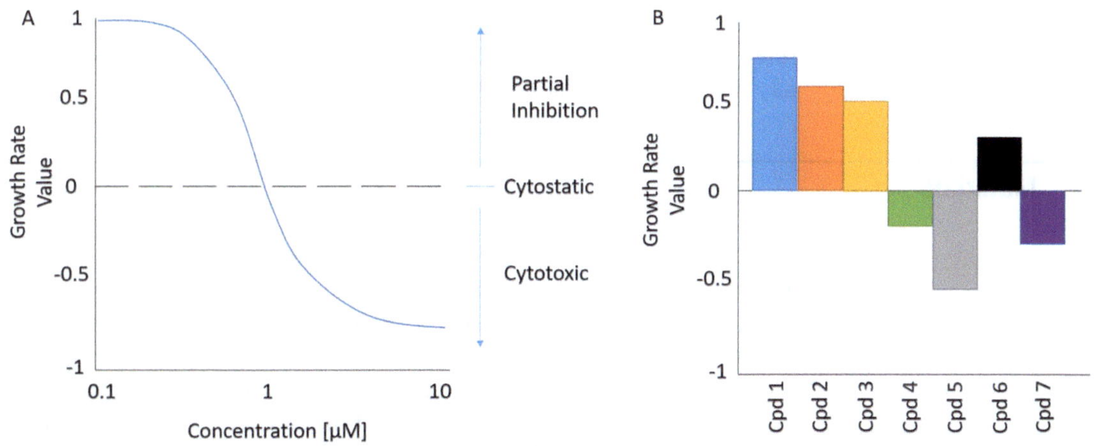

Fig. 1 Graphical illustration of growth rate. (**a**) Example of a dose–response curve for a compound and potential outcomes at differing concentrations. As the concentration of a compound increases it is more likely to exhibit cytotoxic effects and presents graphically as an inverse sigmoid function. (**b**) Comparison of compound effect at a fixed concentration. With this format, it is possible to differentiate between the effect type of each compound and base a selection from the information presented

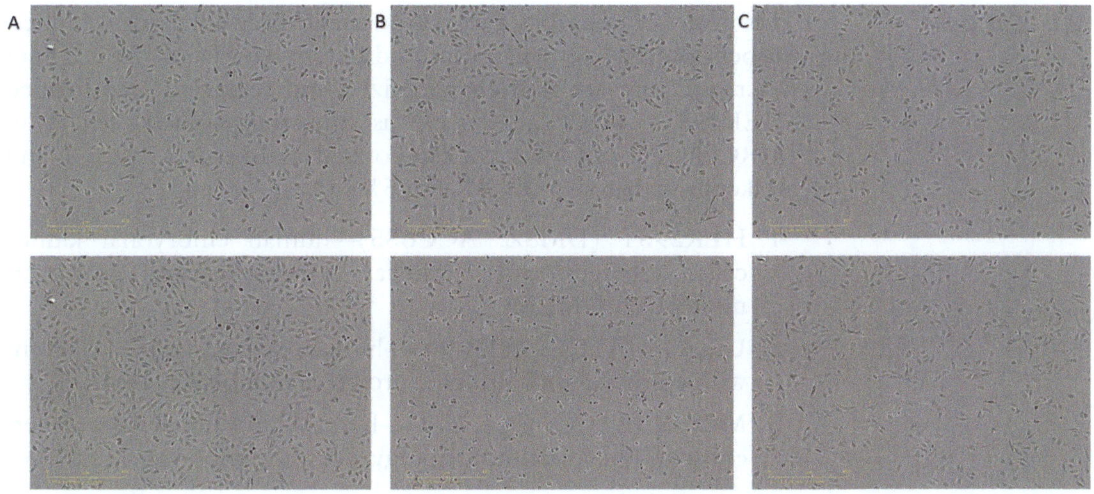

Fig. 2 Brightfield images of U2OS cells incubated with DMSO (**a**) Daunorubicin (**b**) and Milciclib (**c**) at 0h (top) and 24h (bottom) showing healthy, cytotoxic, and cytostatic effects, respectively

involve cell seeding, Incucyte image acquisition setup, image analysis and export and growth rate calculation for compound classification and viability characterization. The approach described here is transferable to other systems, but will require adaptations, depending on the instrument used.

It is recommended to run biological replicates for this assay due to the lack of technical duplicates.

Example images from previous experiments are available for reference on the BioImage Archive database (https://www.ebi.ac.uk/biostudies/studies/S-BIAD145#) [9].

2 Materials

All solutions must be prepared under sterile conditions and kept so; all work involving live samples requires sterile working. Use cell lines that have not undergone a substantial number of passages to ensure genomic integrity [6]. Cell culture reagents including medium, washing solution, and reagents for cells passaging should be stored at 4 °C and be warmed up to 37 °C prior to use. Compounds should be prepared as aliquots and can be stored sealed at room temperature (RT) for approximately 3–6 months. Try to avoid freeze–thaw cycles [7] and ensure that aliquots are soluble at the stored concentration and are not precipitated prior to use. For longer term storage, stock solutions should be stored at − 20 °C.

2.1 Cell Lines

This protocol describes a workflow using three adherent cell lines. Suspension cell lines will require a different workflow. The cell lines described here are well-characterized and represent transformed (HEK293T, U2OS) as well as non-transformed cell lines (MRC-9). All cell lines are categorized as Biosafety Level 1 and are kept incubated at 37 °C with 5% CO_2.

1. HEK293T (DMSZ: ACC635)—human embryonal kidney cells, adherent and epithelial in morphology, grown as monolayer, DMEM.
2. U-2 OS (ATCC: HTB-96)—human osteosarcoma, adherent with epithelial morphology, grown as monolayer, DMEM.
3. MRC-9 (ATCC: CCL-212)—human lung fibroblasts, adherent with fibroblast morphology, grown as monolayer, EMEM.

2.2 Reagents and Perishables

As mentioned previously in Subheading 2.1, the medium listed here is specific to the cell lines we present as an example. Please ensure you use correct media and supplementation for your cell lines of choice.

1. Dulbecco's modified Eagle's medium (DMEM) (Thermo Fisher: 11965084): supplement with 10% fetal bovine serum (Thermo Fisher: 26140079) and 1% penicillin–streptomycin (Thermo Fisher: 15140122).
2. Eagle's minimum essential medium (EMEM) (ATCC:30-2003): supplement with 10% fetal bovine serum (Thermo Fisher: 26140079) and 1% penicillin–streptomycin (Thermo Fisher: 15140122).

3. Cell washing solution: Dulbecco's phosphate-buffered saline 1× (DPBS): without calcium or magnesium (Thermo Fisher: 14190144).
4. Dissociation reagent: trypsin-EDTA 1× (0.05%) (Thermo Fisher: 15400054).
5. 0.4% trypan blue solution (Thermo Fisher: 15250061).
6. Cell culture flask: T75 is sufficient (75 cm^2/250 mL) (Greiner: 658170).
7. Serological pipettes (Eppendorf: 0030127722) and PIPET-BOY (Integra 155017).
8. Mechanical pipettes (Eppendorf: 3123000918).
9. Greiner Bio-One 384-well standard CELLSTAR polystyrene microplate (Greiner: 781091) or compatible 384 microplate.
10. 15-mL and 50-mL tubes (Greiner: 188261/2120261).
11. Disposable reservoirs (Thermo Fisher: 95128095).

2.3 Reference Compounds

Reference compounds should cover outcomes that include cytotoxicity, cytostaticity, cell lysis/permeabilization, and having no observed effect on cell growth at the concentrations used. Described below are reference compounds commonly used as controls for the experiment. These compounds are tested at 10 μM (0.1% v/v for DMSO) and are distributed evenly throughout the plate. If compounds are dissolved in water, exchange DMSO for PBS/Water.

1. Daunorubicin—positive [10].
2. Staurosporine—positive [11].
3. Digitonin—positive [12].
4. Milciclib—positive [13].
5. Bromosporine—positive [14].
6. JQ1—positive [15].
7. Paclitaxel—positive [16].
8. Torin—positive [17].
9. DMSO—no effect [18].

2.4 Instruments and Related Software

1. Incucyte® SX1, S3 or SX5 Live-Cell Analysis System (Sartorius AG), or similar device.
2. Incucyte® Base Analysis Software.
3. Centrifuge.
4. CO_2 incubator.
5. Light microscope.

6. Cell counting device: automated cell counter (i.e., TC20 Bio-Rad) or hemocytometer (Fisher Scientific).

3 Methods

Carry out all procedures at room temperature and under sterile conditions using aseptic technique unless specified otherwise. Seeding density for different cell lines will need to be established prior to the experiment (*see* **Note 1**).

3.1 Cell Seeding

1. Aspirate medium from cell culture flask by angling the flask and add 5 mL 1× DPBS to the corner of the flask; rinse cells with the solution.
2. Aspirate DPBS from the flask and add 3 mL of 1× trypsin-EDTA; return flask to the incubator for 1–2 min (dependent on cell line).
3. Remove flask from incubator and determine the detachment of cells via light microscopy.
4. Once the cells are detached, add 7 mL of culture medium to neutralize trypsin activity and transfer dilution to a 15-mL tube.
5. Spin down at 200× *g* for 5 min.
6. Aspirate supernatant without disturbing the pellet and resuspend in fresh medium.
7. Create a 1:1 dilution of cell suspension and 0.4% trypan blue (10 μL of cells in 10 μL trypan blue), and add 10 μL to the counting slide and insert into the automated cell counter (or count using a hemocytometer).
8. Note the cell number and viability (viability above 95% is acceptable). Create a cell suspension for a desired cell number per well in a chosen volume resulting in a confluency value of approximately 40%. Cell seeding density (cell/mL) is calculated via the formula as follows:

   $$(1000/\text{volume per well}) \times \text{cells per well.}$$

   For example, seeding 1500 cells per well in 50 μL: (1000/50) × 1500 = 30,000 cells/mL solution.
9. Cell numbers for cell lines described above consist of the following:
   (a) HEK293T—1750 cells/well.
   (b) U2OS—1500 cells/well.
   (c) MRC-9—1250 cells/well.

10. Add × μL of your cell solution (volume calculated from **step 8**) to each well of the plate (avoiding the outer perimeter, *see* **Note 4)**, allow to rest for up to 20 min at room temperature (*see* **Note 2**).
11. Remove any air bubbles present in the wells as they can interfere with the sedimentation and visualization of the cells (*see* **Note 3**).
12. Place in incubator at 37 °C 5% CO_2 overnight.
13. The following day check your plates and observe the status of the cells. They should be dispersed evenly throughout each well and be ~40% confluent; adherent cells should be adhered to the plate. From here you can set up the Incucyte image acquisition.

3.2 Incucyte Image Acquisition Setup

The following setup is specifically tailored to the Incucyte live cell imaging system; other devices will require adaptation to the protocol.

The Incucyte software is divided into two main parts: acquiring scans and viewing and analysis of the scans. Acquisition allows for vessel configuration, scheduling, image acquisition, and the storage of vessel data into a database. Viewing and analysis involve measuring and assessing the acquired vessel data. Any general user queries regarding the software can be answered by the user manual [8].

1. Open the Incucyte imaging software and enter your credentials.
2. Select **Schedule** to enter the acquisition window.
3. Select **Launch Vessel Wizard**, the **Add Vessel** wizard opens.
4. Select **Scan on Schedule** and click **Next**, the **Create Vessel Page** opens.
5. The **Create or Restore Vessel Page** provides four options as follows:
   (a) **New**—creates an entirely new vessel to scan.
   (b) **Copy Current**—creates a new vessel by copying a vessel from the current schedule.
   (c) **Copy Previous**—creates a new vessel by copying a previously scanned vessel.
   (d) **Add Scans (Restore)**—restores a previously scanned vessel for additional scanning.
6. Under **Create Vessel** select **New**, this will open the **Scan Type** page.

7. On the **Scan Type** page select **Standard**, this will open the **Scan Settings** page.
8. Specify the image channel **Phase** and objective to **10×**. Click **Next** and the **Vessel Type Search** page will open.
9. On the **Vessel Type** window, you will need to select the correct plate; for our assay, we use Greiner 384 Serial Number 781091 (other plate types can be used but ensure they are compatible and the correct plate is selected). Select **Next** and the **Vessel Locator** page opens.
10. **Select a location** in which the plate will be placed, click **Next**, and the Scan Pattern page will open.
11. The **Scan Pattern** page displays a vessel map which will show available wells to be scanned based on the plate you have selected in the **Vessel Type** window. Drag and drop to select the inner 308 wells and specify the number of images per well as 1 (384 plates only allow 1 image per well) (*see* **Note 4**). Once selected, click **Next** to bring up the **Vessel Information** window.
12. **Vessel Information** requires a **Name for the plate**, and enter cell type and passage at your leisure. A plate map can be created manually or imported to define each well with compound and cellular conditions. Click **Next** and the **Analysis Setup** window will appear.
13. Click **Next** on the **Analysis Setup** window, and analysis will be defined later.
14. Set the scanning schedule by dragging the bar across the timeline and set the scan time to 6-h intervals. In the **Stop Scan** section, select after 24 h (time point measurements can be changed to suit the user's preference). Click Next.
15. The **Summary Page** will outline the experimental run, and click **Add to schedule**.

3.3 Incucyte Plate Run

1. Insert the plate into the cassette that was designated earlier (*see* **step 10** in subheading 3.2) to take the blank measurement (*see* **Notes 5, 6, 10** and **11**). The blank is required for calculation of the growth rate (*see* subheading 3.6).
2. Manipulate the schedule via the timeline on the acquisition window to scan the plate within the next 5–10 min. You can see the progress of the scan at the top right of the acquisition window.
3. Once scanned, select the **View** tab and double-click on the scan; check that each well is visible and there are no resolution problems, smears, or scratches (*see* **Notes** 7 and **8**).

4. Remove the plate from the device, add test and reference compounds to the plate to a final concentration of 10 μM. (10 μM is a good benchmark to expect toxicity if it were to occur). Other concentrations may be used depending on the expected potency of your compounds.
5. Return the plate to the Incucyte, and manipulate the schedule via the timeline on the acquisition window to scan the plate within the next 5 min (*see* **Note 9**).
6. Once scanned, select the **View** tab and double-click on your scan, and ensure that each well is visible and there are no resolution problems, smears, scratches, or apparent precipitation of compounds (*see* **Notes 5** and **6**).
7. Allow the plate to run for full 24 h duration.

3.4 Incucyte Plate Analysis

1. Select your plate via the **View** tab.
2. A plate overview will appear with options on the left side of the window. Check each time point to ensure all wells are in focus.
3. Select **Launch Analysis** from the analysis toolbar.
4. Select **Create a New Analysis Definition**.
5. Select **Basic Analyzer**.
6. Select the **Image Channels**, and this assay only uses **Phase**.
7. Select images for the software to train with, and we suggest to include wells that consist of different morphologies while incorporating a negative and positive control wells for good coverage.
8. Define the analysis parameters; this process will involve trial and error to create a fitting mask (*see* **Notes 12** and **13**).

   For HEK293T/U2OS/MRC-9, the following is recommended as follows:

   (a) Hole fill (μm): 100/100/200.
   (b) Adjust size (pixels): −1/−1/−1.
   (c) Area (μm^2): Select Min and insert 100/100/200.
   (d) Eccentricity: Select Min and insert 0.3–0.5/select Min and insert 0.3–0.5/N/A.
9. Select the **Preview All** function to visualize the defined mask for all selected wells. A well-fitting mask outlines cell morphology well while excluding debris, if acceptable click **Next**.
10. On the **Scan Times** and **Wells** window select **all time points and wells**, click **Next**.
11. To Save and Apply the analysis definition input a **Definition Name**, analysis notes are not required. Click **Next**.

12. Review and verify the summary information, and click **Next** to apply the analysis definition. The software will now run the analysis; you can view the progress of this via the status tab.

3.5 Data Export and Analysis

1. After the vessel data has been analyzed, the vessel window will now have the associated analysis masks that were applied.
2. Open the vessel and click the **Graph Metrics** icon to open the Graph Metrics window.
3. Select the **Confluence (%)** metric from the metrics pane.
4. Select all time points and wells you measured. From here you can select as follows:
   (a) **Microplate Graph**—Shows an overview of every well analyzed and their confluency against time point.
   (b) **Graph**—each well is plotted on a standard xy plot.
   (c) **Export**—export the values to a third-party software.
5. Select **Export** and specify the format in which the data should be exported via the layout and destination can be specified through three options as follows:
   (a) **Clipboard**—to be able to manually paste into a third-party software program such as excel.
   (b) **All scans in one file**—specify where the scans should be stored as a combined file.
   (c) **Each scan in a separate file**—specify where each scan should be stored as individual files.
   (d) Select other options under **Other Options**.

3.6 Growth Rate Calculation

Determining the growth rate is calculated according to the formula [5]:

$$\mathrm{GR}(c) = 2^{\frac{log2(x(c)/x_0)}{log2(x_{ctrl}/x_0)}} - 1$$

$x(c)$ is the treated cell count.

x_{ctrl} is the control cell count.

x_0 is the cell count prior to treatment (Blank).

The workflow and an example of how to calculate this value are illustrated below:

Blank value—12.

After 24 h:

Test compound—13.

Negative control (each well)—25, 26, 27, 28, 29.

To calculate the growth rate:

1. Calculate the negative control average confluency value for the chosen time point.

$$(25 + 26 + 27 + 28 + 29)/5 = 27.$$

2. Take the test compound, blank and negative control confluency values, and input them into the calculation below:

$$\mathrm{GR}(c) = 2^{\frac{log2(x(c)/x_0)}{log2(x_{ctrl}/x_0)}} - 1$$

$x(c)$ is the treated cell count—this is the compound confluency value.

x_{ctrl} is the control cell count—this is the average negative control confluency value.

x_0 is the cell count prior to treatment—this is the blank confluency value.

3. This should be the base formula with the relevant values as follows:

$$\mathrm{GR}(c) = 2^{\frac{log2(13/12)}{log2(27/12)}} - 1$$

13 is the confluency value for the test compound.

12 is the confluency value for the blank measurement.

27 is the confluency value for the negative control.

4. Calculate the first set of brackets. Note - values have been rounded to three significant figures for simplicity.

$$\mathrm{GR}(c) = 2^{\frac{log2(1.08)}{log2(2.25)}} - 1$$

5. Find the log base 2 (log2) of the result.

$$\mathrm{GR}(c) = 2^{\frac{log2(1.08)}{log2(2.25)}} - 1$$

$$\mathrm{GR}(c) = 2^{\frac{0.111}{1.17}} - 1$$

6. Calculate the remaining values.

$$\mathrm{GR}(c) = 2^{0.0949} - 1.$$

7. Finish the calculation.

$$\mathrm{GR}(c) = 0.0680.$$

To minimize the risk of error, it is recommended to use third-party software for calculation of the growth rate, illustrated below is an excel formula that can be used for calculation:

$$= (2\ ((\mathrm{LOG}(((x(c)/x_o), 2)/\mathrm{LOG}((x_{\mathrm{ctrl}}/x_o), 2)))) - 1.$$

$x(c)$ is the treated cell count—this is the compound value.

x_{ctrl} is the control cell count—this is the average negative control value.

x_0 is the cell count prior to treatment—this is the blank value.

4 Notes

1. Different cell lines will require individual optimization to determine a good seeding confluency; for example, we find that most adherent cells work well between 1500 and 2000 cells per well in a 384-well plate. It is important to have sufficient cells to influence growth but not too many so that the initial confluency value is too high.
2. Allowing the plate to sit at room temperature for 20min minimizes the impact of convection currents when moved to the incubator, keeping the cells from aggregating at the edge of the wells.
3. When pipetting your plates, avoid creating bubbles as much as possible, as they can interfere with the attachment and distribution of your cells. If bubbles are formed, it is possible to use a so-called debubbler to remove them. This resembles a laboratory wash, a nozzle inserted into an empty plastic bottle which you are able to squeeze to expel liquid for cleaning purposes (as an example Thermo: 2401-0125). When the bottle is empty and squeezed, air is expelled and will remove any bubbles that have formed within the wells.
4. To avoid the so-called edge-effect, do not use the outer perimeter of the plate, and it can be used as protection by being filled with DPBS or media.
5. When inserting your plate into the Incucyte cassette, do so using a slide motion. Do not press the plate down into the cassette as this can damage the motor of the drawer.
6. Use fresh plates or those that have been sealed previously. Plates that have been standing around can accumulate dust and other debris that will interfere with the analysis software.
7. Avoid touching or abrasively wiping the bottom of the microplate as much as feasibly possible. It can result in smearing and scratches which are detected by the Incucyte optics and analysis software, leading to unreliable confluency values.

8. The Incucyte may have a focusing problem from time to time. If your scan appears blurry, simply rescan at a later time, it should rectify itself.
9. While the Incucyte will display a warning when scheduling additional experiments too close to one another, it is advised to leave at least 5 min before each scan regardless. This ensures that different scans will not overlap and cause a loss of data.
10. The Incucyte will generate heat while working; this is unavoidable. This temperature increase can have an impact on your assay. It is often recommended to set the temperature of the incubator your Incucyte is housed in to 0.5°C lower than 37°C to offset this change.
11. Condensation can build up on the plate's lid when placed into the Incucyte. It is recommended to allow the plate to acclimatize for up to 10min before beginning a scan.
12. Analysis parameters will be cell line specific; different morphologies will dictate the criteria you set.
13. While the parameters here will minimize most unwanted debris from being detected, be aware some compounds can cause a significant increase in confluency values despite causing cell death. An example would be digitonin, a non-ionic detergent that permeabilizes the membrane and causes leakage of cytosolic constituents resulting in a haze effect. It is recommended to assess each image to identify whether the confluency value can be trusted.

Acknowledgement

The authors are grateful for support by the Structural Genomics Consortium (SGC), a registered charity (No: 1097737) that receives funds from Bayer AG, Boehringer Ingelheim, Bristol Myers Squibb, Genentech, Genome Canada through Ontario Genomics Institute, Janssen, Merck KGaA, Pfizer and Takeda and by the German Cancer Research Center DKTK and the Frankfurt Cancer Institute (FCI). This project has received funding from the Innovative Medicines Initiative 2 Joint Undertaking (JU) under grant agreement No 875510. The JU receives support from the European Union's Horizon 2020 research and innovation program and EFPIA and Ontario Institute for Cancer Research, Royal Institution for the Advancement of Learning McGill University, Kungliga Tekniska Hoegskolan, Diamond Light Source Limited. Disclaimer: This communication reflects the views of the authors, and the JU is not liable for any use that may be made of the information contained herein. A.T. is supported by the Deutsche Forschungsgemeinschaft (DFG, German Research Foundation) –

grant number 259130777 (SFB1177). We would also like to thank Dr. Alexandra Stolz and her team for assistance with experimental setup and image optimization.

References

1. Tjaden A, Chaikuad A, Kowarz E, Marschalek R, Knapp S, Schröder M, Müller S (2022) Image based annotation of chemogenomic libraries for phenotypic screening. Molecules 27:1439
2. Müller S et al (2022) Target 2035 – update on the quest for a probe for every protein. RSC Med Chem 13:13–21
3. Moffat JG, Vincent F, Lee JA, Eder J, Prunotto M (2017) Opportunities and challenges in phenotypic drug discovery: an industry perspective. Nat Rev Drug Discov 16:531–543
4. Wagner BK, Schreiber SL (2016) The power of sophisticated phenotypic screening and modern mechanism-of-action methods. Cell Chem Biol 23:3–9
5. Hafner J, Niepel M, Chung M, Sorger PK (2016) Growth rate inhibition metrics correct for confounders in measuring sensitivity to cancer drugs. Nat Methods 13:521–527
6. Hughes P, Marshall D, Reid Y, Parkes H, Gelber C (2018) The costs of using unauthenticated, over-passaged cell lines: how much more data do we need? BioTechniques 43. https://www.future-science.com/doi/10.2144/000112598
7. Kozikowski BA et al (2003) The effect of room-temperature storage on the stability of compounds in DMSO. J Biomol Screen 8: 205–209
8. Sartorius (2022) Incucyte live-cell analysis systems user manual. https://www.sartorius.com/download/1087348/incucyte-live-cell-analysis-systems-user-manual-en-l-8000-04-1%2D%2Ddata.pdf. Accessed 17 June 2022
9. EMBL-EBI (2022) BioImage Archive https://www.ebi.ac.uk/bioimage-archive/. Accessed 4 July 2022
10. Al-Aamri H et al (2019) Time dependent response of daunorubicin on cytotoxicity, cell cycle and DNA repair in acute lymphoblastic leukaemia. BMC Cancer 19. https://doi.org/10.1186/s12885-019-5377-y
11. Bruno S et al (1992) Different effects of staurosporine, an inhibitor of protein kinases, on the cell cycle and chromatin structure of normal and leukemic lymphocytes. Cancer Res 51: 470–473
12. Seina J et al (2001) Cytotoxicity of digitoxin and related cardiac glycosides in human tumor cells. Anti-Cancer Drugs 12:475–483
13. Sanchez-Martinez C, Gelbert LM, Lallena MJ, Dios A (2015) Cyclin dependent kinase (CDK) inhibitors as anticancer drugs. Bioorg Med Chem Lett 25. https://doi.org/10.1016/j.bmcl.2015.05.100
14. Picaud S et al (2016) Promiscuous targeting of bromodomains by bromosporine identifies BET proteins as master regulators of primary transcription response in leukemia. Sci Adv 2. https://doi.org/10.1126/sciadv.1600760
15. Wen N et al (2019) Bromodomain inhibitor JQ1 induces cell cycle arrest and apoptosis of glioma stem cells through VEGF/PI3K/AKT signalling pathway. Int J Oncol 55. https://doi.org/10.3892/ijo.2019.4863
16. Wang TH et al (2000) Paclitaxel-induced cell death. Am Cancer Soc J 88. https://doi.org/10.1002/1097-0142(20000601)88:11<2619::AID-CNCR26>3.0.CO;2-J
17. Francipane MG, Lagasse E (2013) Selective targeting of human colon cancer stem-like cells by the mTOR inhibitor Torin-1. Oncotarget 4. https://doi.org/10.18632/oncotarget.1310
18. Violante GA et al (2002) Evaluation of the cytotoxicity effect of dimethyl sulfoxide (DMSO) on $CaCo_2$/TC/colon tumor cell cultures. Biol Pharmaceut Bull 25:1600–1603

Chapter 7

Plate-Based Screening for DUB Inhibitors

Stephan Scherpe, Aysegul Sapmaz, and Monique P. C. Mulder

Abstract

The evolutionally conserved and abundant post-translational modifier ubiquitin (Ub) is involved in a vast number of cellular processes. Imbalanced ubiquitination is associated with a range of diseases. Consequently, components of the ubiquitylation machinery, such as deubiquitinating enzymes (DUBs) that control the removal of Ub, are emerging as therapeutic targets. Here, we describe a robust assay suitable for small-molecule inhibitor screening. This assay has the potential to drive the development of small-molecule compounds that can selectively target DUBs.

Key words Ubiquitin, Deubiquitinating enzymes, DUBs, Inhibitors, Screening

1 Introduction

Ubiquitination is an important post-translational modification, which plays a crucial role in a wide array of cellular pathways, such as protein homeostasis, DNA repair, and autophagy.[1] Ubiquitination involves the covalent attachment of a small protein, ubiquitin (Ub), to a targeted protein through an enzymatic cascade, which can be reversed through the action of deubiquitinases (DUBs).[2] As a consequence of the critical function of ubiquitination, dysfunction of the enzymes involved in the process contributes to a large number of human diseases.[2] DUBs, for example, have been shown to be involved in a plethora of human diseases such as cancer and neurodegenerative and infectious diseases.[2] Hence, DUBs have become an increasingly important target for drug discovery, with a growing number of novel inhibitors described in recent years.[3]

The identification of inhibitors for DUBs allows for the creation of tools to study the function of these enzymes and possibly the development of treatments for human diseases.[4–6] High-throughput screening to identify such inhibitors, with compound libraries, can be done with a plate-based DUB activity assay (*see*

Daniel Merk and Apirat Chaikuad (eds.), *Chemogenomics: Methods and Protocols*, Methods in Molecular Biology, vol. 2706,
https://doi.org/10.1007/978-1-0716-3397-7_7,

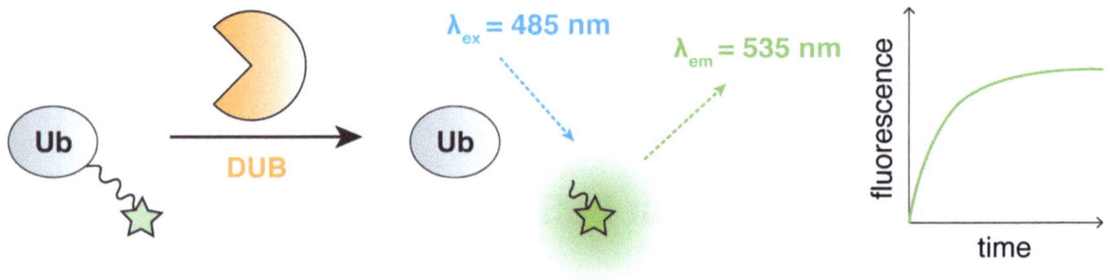

Fig. 1 Schematic representation of the Ub-Rho activity assay. The C-terminal bond between Ub and the fluorophore rhodamine is cleaved by the specific Ub-targeting isopeptidase activity of DUBs. Upon hydrolysis, the fluorescence intensity of the fluorophore increases, thus allowing the real-time monitoring of the activity of DUBs and evaluation of the inhibition effect of small molecules against these enzymes

Fig. 1) using ubiquitin–rhodamine110 (Ub-Rho) as described here [7, 16]. This method provides a fast and reliable assay to investigate the effect of small molecules on the deubiquitinating function of DUBs. Although we here focus on the most prominent family of DUBs, cysteine proteases, the assay can also be used for metallo-based DUBs.[8] Typically, we screen up to 120,000 compounds in 10 days using this methodology and identify novel inhibitors for a diverse set of DUBs. After the initial identification of assay inhibitors ("hits"), these need further investigation to verify the interaction and improve the properties. This is achieved by exploring the structure–activity relationship between the inhibitor hits and the enzyme to increase potency and selectivity, as well as conducting pharmacokinetic studies.[6, 9, 10]

2 Materials

1. Tris/NaCl 10× buffer: 500 mM Tris–HCl, pH 7.5, 1 M NaCl. Weight 78.8 g Tris–HCl and 58.4 g NaCl and transfer into a 1-L beaker with 500 mL of water (*see* **Note 1**). Dissolve by mixing and adjust pH to 7.5 with HCl (*see* **Note 2**). Fill up to 1 L and store at room temperature.
2. 1 M TCEP–HCl, pH 7.0: weight 2.87 g of TCEP–HCl and transfer into a small beaker. Add 8 mL of water, and mix and adjust pH to 7.5 with NaOH. Fill up to 10 mL, divide into 1 mL aliquots, and store at −20 °C. HCl (*see* **Note 3**).
3. Assay buffer: 50 mM Tris–HCl, 100 mM NaCl, 1 mM TCEL–HCl, 1 mg/mL CHAPS (3-[(3-cholamidopropyl) dimethylammonio]-1-propanesulfonate), and 0.5 mg/mL BGG (bovine gamma globulin). Prepare freshly using deionized water and filter (0.45 μm) before use.

4. DUB of interest: concentrated protein (typically in the range of 0.1–10 μM) in buffer, stored in aliquots at −80 °C (*see* **Note 4**).
5. Dimethyl sulfoxide (DMSO).
6. Ubiquitin–rhodamine (Ub-Rho). N-terminal rhodamine-labeled Ub produced by solid-phase peptide synthesis (SPPS), and 1 mM dissolved in DMSO and stored at −20 °C (*see* **Note 5**).
7. 1 M iodoacetamide: Dissolve 185 mg of iodoacetamide in 1 mL of DMSO.
8. Reference compound from literature, dissolved in DMSO; advised concentration 1–100 mM (*optional*).
9. Small-molecule library (*see* **Note 6**).
10. Pipettes for volumes between 1 and 1000 μL.
11. Multichannel pipette for volume between 1 and 20 μL.
12. Eppendorf tubes, 1.5 mL.
13. Falcon tubes, 15 mL or 50 mL.
14. Microplates (*see* **Note 7**).
15. Microplate dispenser (*see* **Note 8**).
16. Centrifuge for microplates.
17. Fluorescence plate reader, with a 485 nm excitation filter (λ_{ex} 485 nm) and emission filters capable of measuring at 535 nm (λ_{em} 535 nm).

3 Methods

3.1 Pre-screening Setup

1. To determine workable DUB concentration for the assay: Dilute DUB enzyme of interest into assay buffer at different concentrations (typically eight concentrations between 0.01 nM and 10 μM: 0.01 nM, 0.1 nM, 1 nM, 10 nM, 100 nM, 100 nM, 1 μM, and 10 μM). Add DUB into assay buffer to obtain 50 μL of 4× final assay concentrations of the highest DUB to be tested (e.g., 40 μM) in a 1.5-mL Eppendorf tube. Dilute 5 μL of that DUB solution into 45 μL assay buffer to obtain the second highest DUB concentration to be tested, and repeat to obtain all desired DUB dilutions in 4× final assay concentration (*see* **Note 9**).
2. For one 384-well plate assay (*see* **Note 7**): Dilute 40 μL of 1 mM substrate (Ub-Rho) into 1960 μL assay buffer in a 15-mL Falcon tube to obtain 2 mL of a 20 μM solution.
3. Pipette 1 μL of DMSO into three wells of a 384-well plate, for each enzyme concentration negative control (*see* **Note 10**).

4. Add 1 μL of 1 M iodoacetamide solution into three wells for each of the enzyme concentrations (positive control).
5. Using a multichannel pipette, add 9 μL of assay buffer to all used wells and mix (*see* **Note 11**).
6. Add 5 μL of DUB solution to three negative and three positive control wells for each concentration and mix (*see* **Note 11**).
7. Incubate plate at room temperature for 1 h.
8. Add 5 μL of Ub-Rho substrate to each well and mix (*see* **Note 11**).
9. Centrifuge the plate at 500× *g* for 4 min.
10. Read the fluorescence continuously to measure the slope of the increasing fluorescence intensity (*optional*) *or* incubate the plate at room temperature for 30 min and then read the fluorescence intensity of the cleaved rhodamine using a microplate reader at λ_{ex} 485 nm and λ_{em} 535 nm.
11. Calculate the Z′ score for each concentration. A reliable assay should have a Z′ of 0.5–1.[11] At least three positive and negative controls must be included in each screening plate (*see* **Note 10**).

$$Z' = 1 - \frac{3\sigma_p + 3\sigma_n}{\left|\mu_p - \mu_n\right|}$$

where

σ_p/σ_n: standard deviation of fluorescence intensity signal (*optional* of the slopes of the signal) of positive/negative control.

μ_p/μ_n: mean of fluorescence intensity signal of positive/negative control.

3.2 Single-Point Inhibitor Screen

1. Centrifuge library plates at 500× *g* for 4 min (*see* **Note 12**).
2. For one 384-well plate assay (*see* **Note 7**): Dilute DUB enzyme of interest into assay buffer for desired concentration (4× final assay concentration) to obtain 2 mL, in a 15-mL Falcon tube.
3. Dilute 40 μL of 1 mM substrate (Ub-Rho) into 1960 μL assay buffer in a 15-mL Falcon tube to obtain 2 mL of a 20 μM solution.
4. Transfer library compounds (DMSO solution) to a mixing plate in order to pre-dilute compounds with DMSO accordingly (*optional*).
5. Transfer 1 μL of individual library compounds at a concentration of 2 mM into separate wells of a 384-well assay plate, for a final assay concentration of 100 μM.

6. Add 1 μL of DMSO into three wells (negative control) (*see* **Note 10**).
7. Add 1 μL of 1 M iodoacetamide solution into three wells (positive control).
8. Using a multichannel pipette, add 9 μL of assay buffer to all wells and mix (*see* **Note 11**).
9. Dispense 5 μL of DUB solution to all wells and mix (*see* **Note 11**).
10. Incubate plate at room temperature for 1 h.
11. Add 5 μL of Ub-Rho substrate to all wells and mix (*see* **Note 11**).
12. Centrifuge the plate at 500× *g* for 4 min.
13. Read the fluorescence continuously to measure the slope of the increasing fluorescence intensity (*optional*) *or* incubate the plate at room temperature for 30 min and then read the fluorescence intensity of the cleaved rhodamine using a microplate reader at λ_{ex} 485 nm and λ_{em} 535 nm.
14. Calculate the Z′ score for each plate.
15. Calculate the inhibition of each library compound at the measured concentration.

$$\text{Inhibition (Compound } A\text{) [\%]} = 100\,\frac{A - \mu_p}{\mu_n - \mu_p}$$

where

A: fluorescence intensity signal of a library compound.

μ_p/μ_n: mean of fluorescence intensity signal of positive/negative control.

3.3 Hit Verification and Follow-Up

Compounds that have been determined to fulfill an inhibition threshold, typically between 50 and 90%, are to be re-screened by cherry-picking.

1. Centrifuge library plates at 500× *g* for 4 min.
2. For one 384-well plate assay (*see* **Note** 7): Dilute DUB enzyme of interest into assay buffer for desired concentration (4× final assay concentration) to obtain 2 mL, in a 15-mL Falcon tube.
3. Dilute 40 μL of 1 mM substrate (Ub-Rho) into 1960 μL assay buffer in a 15-mL Falcon tube to obtain 2 mL of a 20 μM solution.
4. Transfer selected library compounds (DMSO solution) to a mixing plate in order to pre-dilute compounds with DMSO accordingly (*optional*).

5. Transfer 1 μL of selected library compounds at a concentration of 2 mM as three replicates into separate wells of a 384-well assay plate, for a final assay concentration of 100 μM.
6. Add 1 μL of DMSO into three wells (negative control) (*see* **Note 10**).
7. Add 1 μL of 1 M iodoacetamide solution into three wells (positive control).
8. Using a multichannel pipette, add 9 μL of assay buffer to all wells and mix (*see* **Note 11**).
9. Dispense 5 μL of DUB solution to all wells and mix (*see* **Note 11**).
10. Incubate plate at room temperature for 1 h.
11. Add 5 μL of Ub-Rho substrate to all wells and mix (*see* **Note 11**).
12. Centrifuge the plate at 500× *g* for 4 min.
13. Read the fluorescence continuously to measure the slope of the increasing fluorescence intensity (*optional*) *or* incubate the plate at room temperature for 30 min and then read the fluorescence intensity of the cleaved rhodamine using a microplate reader at λ_{ex} 485 nm and λ_{em} 535 nm.
14. Calculate the inhibition of each library compound at the measured concentration as described in Subheading 3.2 by using the mean value for the measured compounds.
15. Identified hits have to be investigated further. To this end, hit compounds are to be obtained and measured in the assay at different concentrations, to establish a dose–response relationship and to calculate the IC_{50}. Hits can then be further validated with orthogonal assays, especially Ub-based activity-based probes, such as Ub–vinyl methyl ester (Ub-VME) and Ub–propargylamine (Ub-PA), and biophysical methods, including isothermal titration calorimetry and NMR. [6, 12–15] The selectivity of hits can be investigated with DUB panels and improved through medicinal chemistry (*see* Ref. [16]).

4 Notes

1. Mixing Tris–HCl into water helps to dissolve it easier. Fill the beaker with water and stir while adding Tris, with the help of a stirring bar or a glass rod.
2. Use HCl in high concentration (10 M) to bring the pH close to the desired value. For the fine adjustment, switch to a lower HCl concentration (1 M) and add slowly.

3. TCEP, or other reducing agents such as dithiothreitol or cysteine, is used to prevent the oxidation of the active-site cysteine. TCEP solution can be stored at −20 °C for months. In order to avoid degradation, it is recommended to aliquot it into smaller volumes and abstain from freeze/thawing.
4. The activity of enzymes, such as DUBs, can often be preserved for some time when stored at −80 °C. Avoid freeze/thawing by aliquoting the DUB into smaller volumes.
5. SPPS-produced Ub-Rho is commercially available.[7] Ub-Rho can be solved in DMSO and stored at −20 °C in aliquots, to avoid degradation through freeze/thawing. In our laboratory, Ub-Rho is typically stored as 1 mM stock solutions and 1000× excess of the substrate to the enzyme is typically used.
6. The choice of libraries often depends on availability in academic settings. DMSO-based libraries of compounds optimized on criteria for small-molecule inhibitors, such as Lipinski's rule of 5, are of great value.[17]
7. The assay can be performed in different plate formats, such as 96, 384, and 1536, and the volume has to be adjusted to the working volume range of the plate. The plates should be made from low-fluorescent material such as polystyrene.
8. Although hand-held pipettes, especially multichannel pipettes, are a useful alternative, an automated system to dispense liquid into the plate can reduce the assay time, as well as man-made errors.
9. The deubiquitinase activity varies among DUB, and the enzyme concentration should be optimized before the screening by titrating different concentrations of DUB in the assay. In our laboratory, DUB enzymes are usually used in the Ub-Rho assay at concentrations of 0.1–10 nM.
10. Typically, only 320 wells are used in a 384-well plate, and the first two and last two columns are reserved for controls to increase confidence in the data.
11. Mixing by pipetting up and down, slowly, five times when using a tip-based dispensing, or by using an orbital shaker when using tip-free dispensing.
12. Compound libraries can be stored at −20 °C to improve shelf life. Frozen libraries have to be allowed to warm up to room temperature before use. The lids of libraries are not to be removed before they have reached room temperature, otherwise the hygroscopic DMSO will quickly absorb water from the atmosphere.

Acknowledgements

This work was supported by the EU/EFPIA/OICR/McGill/KTH/Diamond Innovative Medicines Initiative 2 Joint Undertaking (EUbOPEN Grant No. 875510) and NWO (VIDI Grant VI. 213.110 to M.P.C.M).

References

1. Komander D, Rape M (2012) The ubiquitin code. Ann Rev Biochem 81. https://doi.org/10.1146/annurev-biochem-060310-170328
2. Harrigan JA, Jacq X, Martin NM, Jackson SP (2018) Deubiquitylating enzymes and drug discovery: emerging opportunities. Nat Rev Drug Discov 17. https://doi.org/10.1038/nrd.2017.152
3. Lange SM, Armstrong LA, Kulathu Y (2022) Deubiquitinases: from mechanisms to their inhibition by small molecules. Mol Cell 82. https://doi.org/10.1016/j.molcel.2021.10.027
4. Kooij R, Liu S, Sapmaz A et al (2020) Small-molecule activity-based probe for monitoring ubiquitin C-terminal hydrolase L1 (UCHL1) activity in live cells and zebrafish embryos. J Am Chem Soc 142. https://doi.org/10.1021/jacs.0c07726
5. Panyain N, Godinat A, Lanyon-Hogg T et al (2020) Discovery of a potent and selective covalent inhibitor and activity-based probe for the deubiquitylating enzyme UCHL1, with antifibrotic activity. J Am Chem Soc 142. https://doi.org/10.1021/jacs.0c04527
6. Schauer NJ, Magin RS, Liu X et al (2020) Advances in discovering deubiquitinating enzyme (DUB) inhibitors. J Med Chem 63. https://doi.org/10.1021/acs.jmedchem.9b01138
7. Hassiepen U, Eidhoff U, Meder G et al (2007) A sensitive fluorescence intensity assay for deubiquitinating proteases using ubiquitin-rhodamine110-glycine as substrate. Anal Biochem 371. https://doi.org/10.1016/j.ab.2007.07.034
8. Hameed DS, Sapmaz A, Burggraaff L et al (2019) Development of ubiquitin-based probe for metalloprotease deubiquitinases. Angewandte Chemie International Edition 58. https://doi.org/10.1002/anie.201906790
9. Wang X, D'Arcy P, Caulfield TR et al (2015) Synthesis and evaluation of derivatives of the proteasome deubiquitinase inhibitor b-AP15. Chem Biol Drug Design 86. https://doi.org/10.1111/cbdd.12571
10. Lamberto I, Liu X, Seo H-S et al (2017) Structure-guided development of a potent and selective non-covalent active-site inhibitor of USP7. Cell Chem Biol 24. https://doi.org/10.1016/j.chembiol.2017.09.003
11. Zhang J-H, Chung TDY, Oldenburg KR (1999) A simple statistical parameter for use in evaluation and validation of high throughput screening assays. SLAS Discov 4. https://doi.org/10.1177/108705719900400206
12. Borodovsky A, Ovaa H, Meester WJN et al (2005) Small-molecule inhibitors and probes for ubiquitin- and ubiquitin-like-specific proteases. ChemBioChem 6. https://doi.org/10.1002/cbic.200400236
13. Cho J, Park J, Kim EE, Song EJ (2020) Assay systems for profiling deubiquitinating activity. Int J Mol Sci 21. https://doi.org/10.3390/ijms21165638
14. Ekkebus R, van Kasteren SI, Kulathu Y et al (2013) On terminal alkynes that can react with active-site cysteine nucleophiles in proteases. J Am Chem Soc 135. https://doi.org/10.1021/ja309802n
15. de Jong A, Merkx R, Berlin I et al (2012) Ubiquitin-based probes prepared by total synthesis to profile the activity of deubiquitinating enzymes. ChemBioChem 13. https://doi.org/10.1002/cbic.201200497
16. Varca AC, Casalena D, Chan WC et al (2021) Identification and validation of selective deubiquitinase inhibitors. Cell Chem Biol 28. https://doi.org/10.1016/j.chembiol.2021.05.012
17. Lipinski CA, Lombardo F, Dominy BW, Feeney PJ (2001) Experimental and computational approaches to estimate solubility and permeability in drug discovery and development settings 1PII of original article: S0169-409X(96)00423-1. The article was originally published in Advanced Drug Delivery Reviews 23 (1997) 3–25. 1. Adv Drug Deliv Rev 46. https://doi.org/10.1016/S0169-409X(00)00129-0

Chapter 8

NanoBRET™ Live-Cell Kinase Selectivity Profiling Adapted for High-Throughput Screening

Amanda N. Nieman, Kaitlin K. Dunn Hoffman, Elizabeth R. Dominguez, Jennifer Wilkinson, James D. Vasta, Matthew B. Robers, and Ngan Lam

Abstract

Kinases represent one of the most therapeutically tractable targets for drug discovery in the twenty-first century. However, confirming engagement and achieving intracellular kinase selectivity for small-molecule kinase inhibitors can represent noteworthy challenges. The NanoBRET™ platform enables broad-spectrum live-cell kinase selectivity profiling in most laboratory settings, without advanced instrumentation or expertise. However, the prototype workflow for this selectivity profiling is currently limited to manual liquid handling and 96-well plates. Herein, we describe a scalable workflow with automation and acoustic dispensing, thus dramatically improving the throughput. Such adaptations enable profiling of larger compound sets against 192 full-length protein kinases in live cells, with statistical robustness supporting quantitative analysis.

Key words BRET, NanoBRET™, Kinase profiling, High-Throughput Screening (HTS), Live-cell target engagement

1 Introduction

With over 500 described members, kinases are among the most diverse class of therapeutic targets in cancer, inflammation, and central nervous system disease. Fortunately, these enzymes are tractable targets of small-molecule drugs owing to the generally accessible ATP co-substrate binding pocket present in the majority of the kinome. However, the ATP pocket is highly conserved among kinases, and therefore, achieving kinase inhibitor selectivity represents a challenge in drug discovery. Investigating the binding properties of novel therapeutic drug candidates is critical for the understanding of structure–activity relationships (SAR), kinetics, and potential off-target effects. To this end, numerous biochemical and live-cell assays [1] have been developed to assess the binding capabilities of kinase inhibitors. While each format has its benefits,

Daniel Merk and Apirat Chaikuad (eds.), *Chemogenomics: Methods and Protocols*, Methods in Molecular Biology, vol. 2706, https://doi.org/10.1007/978-1-0716-3397-7_8,

the engagement characteristics for kinase inhibitors in each format can vary dramatically. For example, though cell-free assays effectively quantify intrinsic drug binding affinity, live-cell assays can account for several factors—including membrane permeability, protein–protein interactions, and intracellular localization—that may affect the measured potency of compounds and are impossible to simulate in a biochemically defined system. Furthermore, as intracellular ATP concentrations are variable and vastly exceed ATP concentrations used in cell-free assays (often by orders of magnitude), it is now generally recognized that intracellular ATP levels cause unpredictable offsets between live cell and acellular potency measurements [2]. In contrast, cell-based phenotypic assays may lack adequate target resolution compared to that provided by cell-free biochemical assays. Thus, neither cell-free biochemical assays nor live-cell phenotypic assays are ideal approaches to understand kinase inhibitor selectivity in a cellular context.

An ideal method to understand live-cell kinase inhibitor selectivity would combine the quantitative capabilities and target resolution provided by a cell-free biochemical assay with the live-cell context provided by a phenotypic assay, and the NanoBRET™ Target Engagement (TE) method was developed to address this problem [3]. The NanoBRET™ TE method employs bioluminescent resonance energy transfer (BRET) using NanoLuc® luciferase and a cell-permeable tracer molecule developed from a known binding tool compound coupled to a fluorescent tag (Fig. 1) [3]. Since its introduction, NanoBRET™ has been used to determine intracellular kinase inhibitor selectivity [2], quantify inhibitor potency through IC_{50} profiling, and perform kinetic studies of

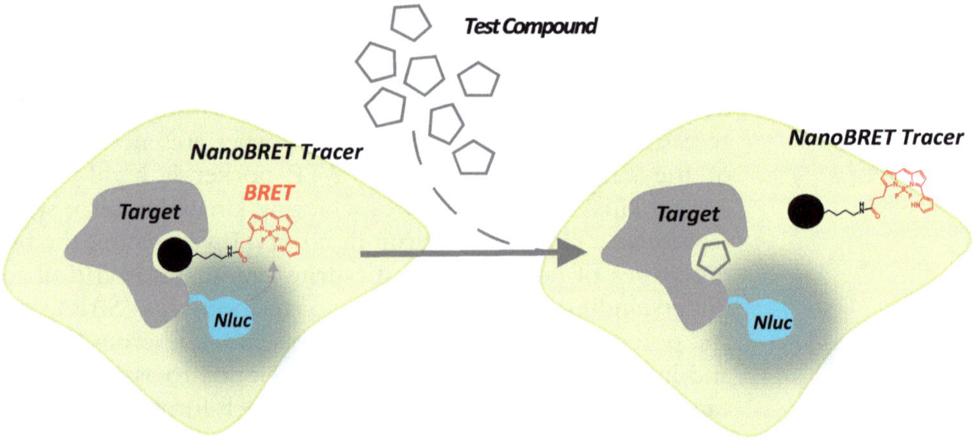

Fig. 1 Schematic of the NanoBRET™ Target Engagement (TE) method. A NanoBRET™ compatible tracer binds to a NanoLuc® fusion protein of interest facilitating proximity-based energy transfer. Displacement of the tracer as a result of compound binding results in the quantifiable reduction of NanoBRET™ signal

residence time for small molecules under non-equilibrium conditions [3, 4].

Screening novel kinase inhibitors against a large number of kinase targets enables a number of unique capabilities, including identification of selective chemical probes for unexplored targets and revealing novel patterns of polypharmacology in both live cells [5] and biochemical environments [6, 7]. A number of biochemical options for kinome-wide screening are available; however, cell-based high-throughput kinase profiling options are limited.

NanoBRET™ TE assays were developed for wide-spectrum target engagement studies across a broad cohort of intracellular kinases. Originally, six tracers were required to quantitatively measure engagement of 178 kinases in cells [2]. This method demonstrated the impact of intracellular ATP levels on drug selectivity for multikinase inhibitors such as dasatinib and crizotinib. A new report was recently published by the same authors that introduced a single tracer, K10, which could be used to study nearly 200 kinases, thus reducing the complexity of the workflow [8]. In this format, only four concentrations of K10 are necessary to accurately measure target engagement of 192 kinases. Despite the simplified workflow using a single tracer, this method has only been validated in 96-well plates using manual pipetting. This dramatically limits the scalability and throughput of this method.

The following chapter outlines the advancements used for miniaturizing and upscaling NanoBRET™ TE assays for 192 kinase targets in 384-well plates, which enables profiling of much larger compound collections in a single experiment. Furthermore, we provide a framework for data analysis and statistical validation that ensures confidence in assay signals and hit determination criteria for each kinase within the kinome diversity panel.

2 Materials

2.1 Optimizing Cell Preparation and Transfection for HTS Kinase Selectivity Profiling

1. Desired cell type (e.g., HEK293, ATCC).
2. Miscellaneous tissue culture reagents and plasticware such as tissue culture-treated flasks, serological pipets, complete media, etc.
3. Diversity panel of 192 NanoLuc®-fused kinase expression vectors driven by mammalian promoter (NanoBRET™ Target Engagement (TE) K192 Kinase Selectivity System, Promega).
4. Control CMV/NanoLuc® Luciferase Expression Vector (Promega).
5. Transfection reagent (e.g., FuGENE® HD, Promega).

6. Transfection carrier DNA (e.g., pGEM-3Z vector, Promega).
7. Opti-MEM™ without phenol red (Gibco).
8. White, tissue culture-treated 384-well plates (Corning).
9. Fetal bovine serum (FBS).
10. Automated liquid dispenser (MultiFlo FX, BioTek).

2.2 Optimization and Z′ Analysis for NanoBRET™ HTS Kinase Selectivity Assays

1. Polypropylene microplates qualified for acoustic dispensing (Echo Qualified Microplates, Beckman Coulter).
2. 400 μM tracer in DMSO (store at −80 °C, avoid free/thaw cycles).
3. 10 mM control compound in DMSO.
4. Dimethyl sulfoxide (DMSO).
5. Acoustic dispenser (Echo, Beckman Coulter).
6. BRET-compatible luminometer equipped with 450 nm (bandpass) and 600 nm (longpass) filters (e.g., GloMax Discover, Promega).
7. NanoBRET™ Nano-Glo® Luciferase Substrate (Promega).
8. Extracellular NanoLuc® Inhibitor (Promega).
9. Automated liquid dispenser (MultiFlo FX, BioTek).

2.3 Validating Performance of Individual Kinase Assay in Measuring Compound Binding Affinity in Concentration–Response Experiments

1. Data analysis software for large datasets (such as JMP).
2. Same materials described in Subheading 2.2 Optimization and Z′ Analysis for NanoBRET™ HTS Kinase Selectivity Assays.

2.4 Evaluating Percent Occupancy of Compounds in Single-Dose Selectivity Profiling

1. Same materials described in Subheading 2.2 Optimization and Z′ Analysis for NanoBRET™ HTS Kinase Selectivity Assays.

3 Methods

3.1 Key Considerations for Adapting NanoBRET™ Kinase Selectivity Assays to HTS

The principles for developing NanoBRET™ TE assays have been discussed previously [4]. Multiple facets of the assay need to be considered during assay validation and verification, which include but are not limited to chemical tracer design, tracer concentration, NanoLuc® tag orientation, and permeabilized or live-cell assay format. A detailed protocol describing assay development in

96-well format has also been published [4]. Building on this prototype, NanoBRET™ TE assays have been developed and validated for targets that span the kinome [8]. These assays are included in the NanoBRET™ TE K192 Kinase Selectivity System that benefits from a single NanoBRET™ Tracer, K10. While this K192 assay simplifies the manual pipetting workflow, this product has only been validated in 96-well plates to date.

Scaling from manual 96-well workflows into an automated workflow requires systematic reoptimization and verification of the profiling system. 384-well plates differ from 96-well plates in their well surface area, evaporation rates, working volume capacity, and surface tension. Therefore, when miniaturizing NanoBRET™ TE assays into a 384-well HTS format, factors such as component ratios, total assay volume, and cell density may need to be reoptimized. During the optimization, outputs such as donor luminescence, homogeneity of the signal across the plate, and control compound performance should be considered.

One of the key steps in adapting the NanoBRET™ TE assay from manual liquid handling in 96-well plates to a 384-well plate format is the utilization of an acoustic liquid handling system, such as an Echo liquid handler from Beckman Coulter. Acoustic dispensing offers quick, versatile, and highly accurate nanoliter transfers, enabling assays to be run in much smaller volumes and with a simplified workflow. Importantly, acoustic dispensing eliminated the need for tracer dilution buffer, a viscous buffer required for manual liquid handling of NanoBRET™ tracers to ensure solubility of the tracer in the aqueous assay medium in concentrated (10× or 20×) solutions. The versatile and simplified tracer handling workflow enabled by acoustic dispensing also allows for increased scalability and potential expansion of the kinase panel using additional kinase NanoBRET™ tracers beyond K10.

As NanoBRET™ TE assays consist of multiple complex steps, it is desirable to break down the workflow into different smaller workflows and optimize them separately to expedite the process. In our laboratory, we optimized cell preparation and transfection concurrently with assay procedure using pre-transfected thaw-and-use cells.

3.2 Cell Preparation and Transfection for HTS Kinase Selectivity Profiling

For robust NanoBRET™ signal, expression levels for the NanoLuc® fusion proteins should generate luminescence sufficiently above reagent background, which is generally defined to be 1000-fold or greater. As higher density plate formats are used, expression levels should be verified in every case. Two main methods exist for the transfection and plating of cells for NanoBRET™ TE assays: bulk transfection prior to cell plating or in-plate transfections. Bulk transfections offer greater consistency between wells and are the preferred method when a smaller number of target proteins are investigated with a larger number of compounds. Additionally, cells from large-scale transfections can be cryopreserved and then plated day of assay (here termed thaw-and-use cells) for multiple runs, which offers significant time savings and additional run-to-run consistency (*See* **Note 1**). In-plate transfections trade well-to-well consistency for ease of use when a large number of target proteins are being investigated with a smaller number of compounds, such as target profiling. Because NanoBRET™ provides a ratiometric output, donor-level variations have minimal effect on NanoBRET™ results if the luminescent donor signal is sufficient for good energy transfer [2, 3]. Variability of in-plate transfections can arise from differing cell densities, passage numbers, or handling between assays, but can be reduced by using a "cells as reagent" product, such as the TransfectNow™ HEK293 cells (Promega).

For both transfection methods, the number of cells plated per well needs to be optimized to achieve sufficient donor luminescence. A balance is required between cell counts high enough to achieve sufficient donor luminescence but low enough to maintain cell health and viability throughout the assay. Target proteins of interest with lower levels of expression may require higher cell densities, while lower cell densities can be used if the NanoLuc® fusion protein is expressed well in cells.

Bulk transfection protocols are typically consistent between assay development and miniaturization as long as the donor signal in the 450 nm bandpass channel is still sufficient as cell numbers decrease in the smaller wells of the 384-well plates. If donor signal is no longer robust enough for quality NanoBRET™ signal and the maximum number of cells are already being plated per well, the DNA concentration of the NanoLuc® fusion protein may need to be increased during transfection. If the expression level of the NanoLuc® fusion protein cannot be increased any further, the assay may not be suitable for miniaturization.

In-plate transfections offer more flexibility with a greater number of targets, but require more optimization through the miniaturization process. DNA concentration, transfection reagent concentration, and volumes of all components need to be optimized to maintain sufficient expression of the NanoLuc® fusion protein such that the donor luminescence signal is at least 1000-fold over machine background to ensure the energy transfer to the acceptor fluorophore can be measured with confidence. Additional optimization may be required if DNA and transfection reagents are dispensed with automated liquid handlers or dispensers to ensure volumes are compatible with available instrumentation.

The following is an example of an optimized transfection protocol that was used to transfect 192 vectors encoding different kinase target proteins in the NanoBRET™ Target Engagement Selectivity System. The resulting donor relative luminescence units (RLUs), shown in Fig. 2 and Table 1, were sufficient for reliable NanoBRET™ Target Engagement assays (protocols for measuring luminescence are provided in the subsequent sections).

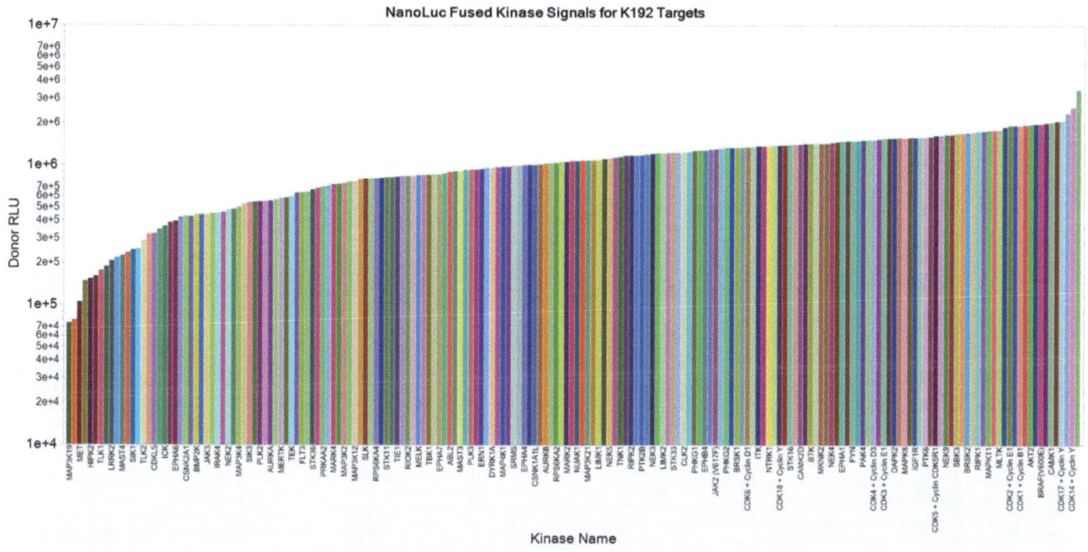

Fig. 2 Donor RLUs resulting from NanoLuc® fusion proteins from the K192 kinase panel. HEK293 cells were transiently transfected with kinases fused to NanoLuc® luciferase at the manufacturer recommend DNA concentrations in technical quadruplets. Average donor RLU is plotted and also listed in Table 1. All donor levels were at least 1000-fold over machine background

1. Split HEK293 cells one day prior to transfection such that cells are 80–90% confluent and in the growth phase on the day of transfection.
2. Prepare DNA for transfection by diluting NanoLuc® fusion vectors in transfection carrier DNA at concentrations verified in assay development at a final total DNA concentration of 20 ng/μL.
3. Dispense 8 μL of Opti-MEM per well using an automated dispenser.
4. Dispense 2 μL of DNA per well using an automated liquid handler.
5. Dilute Fugene HD in Opti-MEM. For each well of a 384-well plate, dilute 0.16 μL Fugene HD in 5 μL of Opti-MEM. Then, dispense diluted Fugene HD solution, 5 μL per well. Allow transfection complexes to form for 30 min (*See* **Note 2**).
6. Trypsinize HEK293 cells that were split one day prior and resuspend at 1.4 × 10^6 cells/mL in Opti-MEM + 4% FBS. The final FBS concentration in the transfection is 1%.
7. Dispense 5 μL of cell suspension per well using an automated dispenser (7 × 10^3 cells/well).
8. Incubate at 37 °C and 5% CO_2 overnight.
9. Proceed to NanoBRET™ assay as described in Subheading 3.3.

Table 1
Conditions and assay performance for individual targets run in the NanoBRET™ HTS kinase selectivity assay

Kinase	[Tracer] nM	Tracer	Mean donor RLU	Assay window	Z prime	MSR	Mean (% occupancy at mean logEC50)	Std dev (% occupancy at mean logEC50)	CV (% occupancy at mean logEC50)	Mean (% occupancy at mean logEC30)	Std dev (% occupancy at mean logEC30)	CV (% occupancy at mean logEC30)
AAK1	25	K10	441600	6.3	0.87	3.6	0.45	0.13	29	0.26	0.14	53
ABL2	1000	K10	895525	2.4	0.9	2.5	0.49	0.07	14	0.30	0.06	21
AKT2	1000	K10	1964625	2.6	0.91	1.7	0.49	0.04	9	0.29	0.06	20
AURKA	25	K10	554500	6.0	0.74	2.9	0.48	0.09	18	0.28	0.08	27
AURKB	25	K10	1029712.5	8.3	0.65	2.5	0.49	0.08	15	0.30	0.07	23
AURKC	25	K10	871862.5	7.5	0.83	2	0.48	0.08	17	0.27	0.11	40
AXL	250	K10	931262.5	2.7	0.81	2.4	0.48	0.07	15	0.29	0.07	24
BMP2K	100	K10	440750	4.8	0.92	3.8	0.48	0.09	19	0.29	0.09	30
BMX	1000	K10	923950	3.4	0.86	2.4	0.50	0.07	13	0.31	0.06	20
BRAF(V600E)	100	K10	1989250	2.7	0.6	2.1	0.49	0.06	12	0.29	0.07	23
BRSK1	250	K10	1338575	5.7	0.81	3.2	0.50	0.11	22	0.31	0.11	37
BRSK2	1000	K10	1711750	17.1	0.71	3.8	0.50	0.12	24	0.31	0.10	34
BTK	1000	K10	1430000	5.9	0.85	2.6	0.50	0.08	15	0.30	0.07	24
CAMK1	1000	K10	2041375	2.3	0.85	3.3	0.51	0.09	17	0.32	0.07	23
CAMK2A	1000	K10	1883375	3.6	0.93	2.6	0.49	0.07	15	0.29	0.06	22
CAMK2D	1000	K10	1418250	3.3	0.8	2.7	0.49	0.08	17	0.29	0.07	24
CDK1 + cyclin B1	1000	K10	1932875	2.5	0.69	2.4	0.50	0.07	14	0.31	0.06	19
CDK2 + cyclin E1	1000	K10	1925875	11.8	0.88	2.5	0.51	0.07	14	0.31	0.06	21
CDK3 + cyclin E1	1000	K10	1552500	5.9	0.84	3.2	0.50	0.09	19	0.30	0.08	26
CDK4 + cyclin D3	1000	K10	1523500	6.9	0.79	2.6	0.49	0.09	19	0.29	0.06	22

CDK5 + cyclin CDK5R1	1000	K10	1639125	10.0	0.92	2.7	0.51	0.08	16	0.32	0.07	24
CDK6 + cyclin D1	1000	K10	1351625	4.0	0.8	2.4	0.49	0.08	15	0.29	0.05	19
CDK7	1000	K10	564737.5	3.4	0.77	2.7	0.48	0.08	16	0.29	0.06	21
CDK9 + cyclin K	1000	K10	816825	4.6	0.86	2.5	0.50	0.07	14	0.30	0.06	21
CDK10 + cyclin L2	1000	K10	805700	2.9	0.39	2	0.49	0.05	11	0.29	0.05	19
CDK14 + cyclin Y	1000	K10	2601875	3.3	0.76	2.6	0.49	0.07	14	0.30	0.06	20
CDK15 + cyclin Y	1000	K10	1085037.5	3.0	0.64	2.9	0.46	0.08	18	0.27	0.08	29
CDK16 + cyclin Y	1000	K10	3507375	5.0	0.88	1.6	0.48	0.05	10	0.28	0.05	16
CDK17 + cyclin Y	1000	K10	2086625	7.3	0.9	1.6	0.48	0.04	8	0.28	0.04	15
CDK18 + cyclin Y	1000	K10	1385875	9.5	0.9	1.9	0.50	0.06	11	0.30	0.04	14
CDK20 + cyclin H	1000	K10	1070862.5	7.6	0.75	2	0.50	0.05	10	0.30	0.06	20
CDKL1	1000	K10	527725	4.3	0.84	2.2	0.49	0.06	13	0.30	0.06	20
CDKL2	1000	K10	728700	8.1	0.89	1.9	0.49	0.07	13	0.29	0.07	22
CDKL3	1000	K10	217850	1.9	0.63	2.9	0.47	0.08	18	0.29	0.08	27
CDKL5	1000	K10	322875	4.2	0.69	2.1	0.48	0.07	15	0.28	0.07	24
CHEK2	1000	K10	1604250	1.9	0.75	3.4	0.49	0.09	18	0.29	0.08	28
CLK1	250	K10	235462.5	5.1	0.79	1.8	0.48	0.05	11	0.29	0.06	22
CLK2	1000	K10	1241962.5	3.6	0.87	2	0.49	0.06	12	0.29	0.05	18
CLK4	100	K10	187287.5	3.5	0.52	6.9	0.42	0.18	43	0.23	0.17	71
CSNK1A1L	1000	K10	1006637.5	2.3	0.75	2.2	0.48	0.03	7	0.29	0.05	16
CSNK1D	1000	K10	969437.5	2.6	0.87	2.2	0.47	0.07	14	0.28	0.06	23
CSNK1G2	1000	K10	2071375	2.4	0.8	2.2	0.49	0.06	12	0.29	0.06	21
CSNK2A1	1000	K10	427975	2.6	0.68	3.5	0.49	0.08	17	0.30	0.08	27

(continued)

Table 1
(continued)

Kinase	[Tracer] nM	Tracer	Mean donor RLU	Assay window	Z prime	MSR	Mean (% occupancy at mean logEC50)	Std dev (% occupancy at mean logEC50)	CV (% occupancy at mean logEC50)	Mean (% occupancy at mean logEC30)	Std dev (% occupancy at mean logEC30)	CV (% occupancy at mean logEC30)
CSNK2A2	100	K10	994987.5	3.6	0.83	6	0.48	0.09	18	0.28	0.12	41
DAPK2	1000	K10	1565375	2.3	0.82	3.8	0.50	0.09	19	0.30	0.08	27
DCLK3	1000	K10	459887.5	2.0	0.62	2.6	0.49	0.07	13	0.29	0.07	24
DYRK1A	1000	K10	957550	2.2	0.66	2.1	0.48	0.06	11	0.29	0.05	18
DYRK1B	1000	K10	1106550	2.6	0.78	2.6	0.47	0.07	15	0.28	0.06	21
EPHA1	1000	K10	1485875	2.5	0.76	2.8	0.46	0.07	16	0.27	0.06	21
EPHA4	1000	K10	1002975	2.4	0.91	2.1	0.47	0.05	11	0.29	0.05	18
EPHA6	1000	K10	395000	3.4	0.71	2.3	0.43	0.05	11	0.26	0.05	19
EPHA7	1000	K10	858725	4.1	0.9	3	0.46	0.09	19	0.28	0.09	32
EPHB1	1000	K10	1523000	2.6	0.81	3.1	0.46	0.09	19	0.27	0.08	31
EPHB4	1000	K10	1279250	2.5	0.78	2.3	0.45	0.07	15	0.27	0.07	26
ERN1	1000	K10	937587.5	2.7	0.84	2	0.47	0.05	11	0.28	0.05	20
ERN2	1000	K10	840175	4.7	0.88	2.2	0.48	0.06	13	0.29	0.06	21
FER	1000	K10	148125	2.3	0.73	3.5	0.48	0.07	15	0.31	0.08	25
FES	1000	K10	1275500	2.4	0.78	2.6	0.50	0.05	10	0.31	0.05	17
FGFR1	100	K10	956650	2.5	0.68	4.3	0.46	0.08	17	0.27	0.07	27
FGFR2	100	K10	1171625	2.7	0.76	2.1	0.46	0.08	18	0.26	0.10	39
FGFR3	250	K10	1953375	2.9	0.87	2.6	0.48	0.09	18	0.28	0.07	24
FGFR4	1000	K10	1414875	2.9	0.78	2.1	0.46	0.07	14	0.27	0.06	22
FLT3	250	K10	634750	3.0	0.9	2.9	0.46	0.09	19	0.27	0.08	31

FYN	1000	K10	1494500	3.0	0.27	3.2	0.48	0.10	20	0.28	0.08	28
GAK	25	K10	1235000	9.3	0.84	2.4	0.47	0.11	24	0.28	0.13	46
HIPK2	1000	K10	152825	2.1	0.29	3.9	0.44	0.08	19	0.26	0.08	30
HIPK3	1000	K10	78066.25	2.2	0.46	9.4	0.41	0.12	29	0.26	0.11	44
HIPK4	100	K10	1034075	3.3	0.28	2.8	0.40	0.25	63	0.24	0.22	95
ICK	1000	K10	364487.5	3.2	0.72	3.2	0.48	0.05	10	0.29	0.04	15
IGF1R	250	K10	1584000	1.9	0.73	3.1	0.46	0.10	22	0.28	0.10	36
IKBKE	100	K10	1226250	5.1	0.6	4	0.47	0.08	18	0.26	0.08	30
INSR	250	K10	1437000	2.2	0.87	3.2	0.46	0.10	22	0.27	0.10	36
IRAK3	25	K10	754050	3.1	0.76	2.6	0.46	0.09	19	0.27	0.08	31
IRAK4	100	K10	454000	4.3	0.64	3.2	0.48	0.11	22	0.28	0.09	33
ITK	100	K10	1367250	2.6	0.84	3.3	0.49	0.13	27	0.29	0.12	40
JAK2 (V617F)	100	K10	1305750	2.7	0.78	3.4	0.48	0.11	24	0.28	0.10	37
JAK3	1000	K10	441925	3.4	0.69	3.1	0.47	0.08	17	0.28	0.08	28
JNK3	1000	K10	543862.5	2.7	0.8	3.3	0.50	0.07	15	0.30	0.07	24
LATS1	100	K10	249837.5	3.8	0.67	3.9	0.48	0.10	21	0.29	0.09	31
LATS2	25	K10	834537.5	3.7	0.79	4.7	0.46	0.13	28	0.29	0.13	45
LCK	1000	K10	1660125	3.1	0.77	2.8	0.47	0.09	20	0.27	0.09	33
LIMK1	1000	K10	1086437.5	4.4	0.93	2.4	0.48	0.07	15	0.29	0.07	26
LIMK2	250	K10	1221500	3.1	0.38	4.9	0.48	0.13	28	0.29	0.15	52
LRRK2	25	K10	206375	2.4	0.62	4.2	0.46	0.11	23	0.29	0.12	43
LTK	100	K10	976400	3.0	0.87	2.5	0.48	0.08	16	0.29	0.08	29
MAP3K2	1000	K10	740000	2.4	0.82	3.5	0.44	0.11	25	0.26	0.10	39
MAP3K3	1000	K10	1391750	2.3	0.88	2.4	0.41	0.05	12	0.25	0.05	20

(continued)

Table 1
(continued)

Kinase	[Tracer] nM	Tracer	Mean donor RLU	Assay window	Z prime	MSR	Mean (% occupancy at mean logEC50)	Std dev (% occupancy at mean logEC50)	CV (% occupancy at mean logEC50)	Mean (% occupancy at mean logEC30)	Std dev (% occupancy at mean logEC30)	CV (% occupancy at mean logEC30)
MAP3K4	1000	K10	499612.5	3.8	0.72	2.7	0.48	0.09	18	0.29	0.08	28
MAP3K9	250	K10	1586250	3.2	0.67	2.8	0.47	0.09	19	0.28	0.10	37
MAP3K10	250	K10	1930500	3.3	0.64	2.4	0.48	0.06	13	0.29	0.07	26
MAP3K11	100	K10	1136125	3.7	0.74	3.1	0.42	0.22	53	0.26	0.17	65
MAP3K12	1000	K10	761512.5	2.6	0.61	2.6	0.42	0.06	15	0.25	0.07	30
MAP3K19	1000	K10	74062.5	2.4	0.36	2.6	0.36	0.19	52	0.23	0.24	105
MAP3K21	1000	K10	1081725	2.1	0.76	2.7	0.45	0.07	14	0.27	0.04	15
MAP4K1	1000	K10	975825	3.3	0.82	2.3	0.47	0.07	16	0.29	0.07	26
MAP4K2	100	K10	1424750	2.9	0.61	4.5	0.49	0.14	28	0.30	0.14	46
MAP4K3	1000	K10	792287.5	4.9	0.79	2.7	0.47	0.09	18	0.28	0.09	30
MAP4K5	1000	K10	797825	2.5	0.87	3.7	0.46	0.09	19	0.28	0.06	22
MAPK4	25	K10	1495625	13.6	0.84	2.7	0.49	0.11	22	0.29	0.10	34
MAPK6	25	K10	428150	10.6	0.66	5	0.41	0.12	30	0.20	0.14	72
MAPK8	100	K10	1579125	3.5	0.64	4.5	0.50	0.09	18	0.29	0.08	27
MAPK9	100	K10	1580125	5.5	0.71	3.6	0.48	0.13	27	0.29	0.11	39
MAPK11	1000	K10	1778250	4.7	0.81	2.2	0.50	0.07	13	0.31	0.06	21
MAPK14	1000	K10	1317375	3.6	0.7	2.6	0.51	0.07	13	0.31	0.06	20
MARK2	1000	K10	1057625	3.2	0.84	3.7	0.48	0.10	21	0.29	0.09	32
MARK4	100	K10	722737.5	3.2	0.78	3	0.48	0.09	18	0.30	0.09	31
MAST3	1000	K10	907675	2.1	0.79	3	0.46	0.08	17	0.28	0.07	27

MAST4	1000	K10	224962.5	2.2	0.71	3.8	0.47	0.08	18	0.28	0.06	23
MELK	1000	K10	845425	2.7	0.68	6	0.43	0.13	30	0.27	0.13	48
MERTK	1000	K10	580787.5	2.3	0.79	2.2	0.48	0.06	13	0.28	0.05	18
MET	1000	K10	104708.75	4.5	0.85	2.3	0.45	0.04	10	0.28	0.04	16
MKNK2	1000	K10	1432625	2.6	0.82	2.8	0.49	0.07	15	0.29	0.07	22
MLTK	250	K10	1791750	2.4	0.67	5.8	0.48	0.13	27	0.29	0.15	51
MUSK	100	K10	482862.5	2.3	0.69	2.3	0.46	0.06	14	0.27	0.07	26
MYLK2	25	K10	1336750	4.4	0.79	3.7	0.49	0.15	30	0.30	0.13	42
MYLK3	250	K10	1647875	5.2	0.64	2.4	0.50	0.08	16	0.29	0.11	36
MYLK4	100	K10	1013600	3.7	0.69	1.8	0.46	0.06	13	0.25	0.08	31
NEK1	1000	K10	451250	3.6	0.83	2.2	0.49	0.07	14	0.30	0.07	24
NEK2	1000	K10	477850	2.8	0.81	3.1	0.47	0.11	23	0.29	0.10	35
NEK3	250	K10	1207250	6.7	0.65	2.7	0.52	0.10	19	0.32	0.11	34
NEK4	1000	K10	1453500	2.5	0.8	2.6	0.46	0.07	16	0.28	0.08	28
NEK5	25	K10	1122000	4.0	0.74	3.8	0.45	0.10	23	0.25	0.12	48
NEK9	250	K10	1656500	4.3	0.97	3.5	0.48	0.11	24	0.29	0.12	40
NEK11	1000	K10	1340375	3.7	0.85	2.3	0.50	0.07	13	0.30	0.07	23
NIM1K	250	K10	1300625	2.9	0.75	4.2	0.48	0.09	19	0.29	0.10	33
NLK	1000	K10	386637.5	2.7	0.79	1.8	0.47	0.06	13	0.27	0.06	23
NTRK1	100	K10	1377000	2.8	0.78	3.2	0.45	0.11	24	0.26	0.10	38
NTRK2	100	K10	1366000	3.0	0.43	3.1	0.44	0.21	47	0.29	0.16	55
NUAK1	25	K10	1072137.5	2.2	0.68	2.3	0.50	0.18	36	0.25	0.11	44
PAK4	1000	K10	1511250	11.9	0.9	2.2	0.50	0.06	12	0.30	0.06	19
PAK6	1000	K10	1788750	2.7	0.71	4.5	0.45	0.12	27	0.28	0.12	42

(continued)

Table 1
(continued)

Kinase	[Tracer] nM	Tracer	Mean donor RLU	Assay window	Z prime	MSR	Mean (% occupancy at mean logEC50)	Std dev (% occupancy at mean logEC50)	CV (% occupancy at mean logEC50)	Mean (% occupancy at mean logEC30)	Std dev (% occupancy at mean logEC30)	CV (% occupancy at mean logEC30)
PHKG1	250	K10	1269250	4.5	0.76	5	0.48	0.11	23	0.28	0.12	43
PHKG2	1000	K10	1330250	2.3	0.73	2.8	0.47	0.06	12	0.27	0.06	23
PKMYT1	1000	K10	711525	2.1	0.6	1.8	0.44	0.04	8	0.26	0.03	13
PLK2	1000	K10	545262.5	2.8	0.64	2.7	0.46	0.09	20	0.27	0.08	28
PLK3	1000	K10	928950	2.4	0.88	3.4	0.46	0.08	18	0.28	0.08	29
PLK4	1000	K10	321537.5	6.2	0.83	3.6	0.47	0.10	22	0.29	0.10	35
PRKAA1	100	K10	581275	6.1	0.79	2.7	0.49	0.08	16	0.30	0.07	25
PRKAA2	100	K10	693725	8.6	0.79	2.8	0.49	0.10	20	0.29	0.10	33
PRKACA	1000	K10	2365625	5.3	0.83	2.9	0.47	0.08	18	0.28	0.08	30
PRKACB	1000	K10	2015125	3.2	0.63	3.1	0.37	0.07	19	0.22	0.07	33
PRKCE	1000	K10	1578750	2.8	0.77	14.4	0.51	0.07	14	0.29	0.07	23
PRKX	100	K10	1384375	2.5	0.33	2.8	0.49	0.12	24	0.30	0.12	40
PTK2	100	K10	904987.5	2.4	0.7	2.1	0.49	0.06	13	0.29	0.05	17
PTK2B	250	K10	1178500	2.3	0.88	3.9	0.49	0.12	24	0.30	0.11	37
PTK6	100	K10	1593500	2.8	0.85	3.1	0.48	0.15	31	0.28	0.15	52
RET	100	K10	1466500	3.7	0.67	3	0.46	0.11	24	0.27	0.11	42
RIOK2	1000	K10	834987.5	6.5	0.76	4.6	0.49	0.09	19	0.30	0.10	33
RIPK1	1000	K10	1757875	2.5	0.73	1.7	0.46	0.04	9	0.27	0.05	20
RIPK2	250	K10	1172687.5	3.8	0.79	3.8	0.49	0.17	34	0.24	0.09	35

RON	1000	K10	681525	3.0	0.8	2.5	0.46	0.07	14	0.28	0.07	26
RPS6KA1	25	K10	1736500	3.8	0.93	2.4	0.47	0.11	23	0.26	0.11	42
RPS6KA2	100	K10	1041500	3.3	0.82	3.1	0.47	0.10	22	0.28	0.09	33
RPS6KA3	25	K10	1248750	2.7	0.79	2.3	0.49	0.08	16	0.29	0.07	24
RPS6KA4	25	K10	800137.5	3.6	0.87	2.7	0.49	0.09	19	0.30	0.08	27
RPS6KA6	100	K10	1564375	2.5	0.81	1.9	0.50	0.07	14	0.30	0.06	20
SBK3	250	K10	1695250	3.1	0.85	2.4	0.48	0.06	13	0.28	0.06	22
SGK1	1000	K10	640912.5	2.9	0.84	2.9	0.48	0.07	15	0.29	0.07	24
SIK1	100	K10	247625	6.2	0.57	2.2	0.52	0.16	31	0.26	0.12	47
SIK2	25	K10	633250	10.0	0.81	2.7	0.46	0.13	27	0.26	0.12	45
SIK3	1000	K10	539475	10.3	0.92	2.2	0.47	0.05	12	0.28	0.05	19
SLK	1000	K10	797700	2.2	0.79	2.4	0.51	0.08	16	0.30	0.07	23
SNRK	1000	K10	1053900	3.1	0.66	3.3	0.49	0.10	20	0.29	0.08	29
SRMS	1000	K10	993025	2.4	0.91	3	0.45	0.07	16	0.27	0.07	27
STK3	1000	K10	1982625	2.6	0.62	3.5	0.49	0.10	20	0.29	0.09	30
STK4	250	K10	1543500	3.0	0.71	3.6	0.49	0.10	20	0.30	0.11	38
STK10	1000	K10	1699750	2.0	0.74	2.4	0.49	0.06	12	0.28	0.07	25
STK11	1000	K10	814925	3.5	0.78	2.6	0.48	0.10	20	0.29	0.09	30
STK16	100	K10	1402875	9.3	0.89	3.1	0.49	0.10	21	0.30	0.09	31
STK17B	1000	K10	425087.5	1.9	0.63	2.9	0.50	0.07	14	0.31	0.05	16
STK32B	1000	K10	1371500	3.5	0.92	4.5	0.50	0.12	23	0.31	0.10	33
STK33	250	K10	1234375	2.6	0.85	2.2	0.48	0.09	18	0.27	0.08	30
STK35	1000	K10	1431000	3.6	0.77	3.5	0.48	0.10	21	0.29	0.09	32
STK36	250	K10	664312.5	5.3	0.68	3.5	0.47	0.08	17	0.27	0.09	34

(continued)

Table 1
(continued)

Kinase	[Tracer] nM	Tracer	Mean donor RLU	Assay window	Z prime	MSR	Mean (% occupancy at mean logEC50)	Std dev (% occupancy at mean logEC50)	CV (% occupancy at mean logEC50)	Mean (% occupancy at mean logEC30)	Std dev (% occupancy at mean logEC30)	CV (% occupancy at mean logEC30)
STK38	1000	K10	1003350	5.8	0.76	3.2	0.48	0.10	21	0.29	0.09	32
STK38L	1000	K10	851400	3.3	0.82	2.9	0.48	0.08	17	0.28	0.07	24
TBK1	100	K10	852762.5	5.4	0.84	3.7	0.49	0.10	21	0.30	0.09	30
TEC	250	K10	1221375	3.3	0.73	5.7	0.49	0.09	18	0.29	0.09	32
TEK	25	K10	595162.5	2.6	0.74	3.4	0.47	0.12	26	0.27	0.10	38
TIE1	250	K10	822850	9.2	0.69	5.5	0.52	0.11	21	0.34	0.13	37
TLK1	1000	K10	176237.5	2.6	0.69	75.2	0.53	0.09	16	0.33	0.10	31
TLK2	1000	K10	286937.5	2.5	0.63	13.2	0.55	0.09	16	0.34	0.10	30
TNK1	25	K10	1147250	5.6	0.36	2.6	0.49	0.09	18	0.29	0.07	24
TNK2	100	K10	1196250	3.1	0.68	2.5	0.44	0.07	16	0.27	0.07	25
TNNI3K	1000	K10	1759875	3.4	0.85	2.1	0.47	0.05	12	0.28	0.06	20
TXK	1000	K10	159675	2.5	0.35	2.9	0.44	0.08	19	0.25	0.09	37
TYRO3	1000	K10	1081375	2.5	0.76	3.2	0.45	0.09	21	0.26	0.08	33
ULK1	100	K10	857737.5	5.9	0.33	2.6	0.50	0.08	17	0.30	0.05	17
ULK2	250	K10	346537.5	3.8	0.6	5.1	0.46	0.11	24	0.28	0.11	40
ULK3	250	K10	1175375	10.8	0.73	7.6	0.47	0.14	29	0.28	0.17	61
WEE1	100	K10	550300	3.1	0.68	3.3	0.43	0.06	14	0.27	0.06	21
WEE2	1000	K10	1494250	2.4	0.84	2.8	0.51	0.09	18	0.30	0.09	29

3.3 Optimization and Z′ Analysis for NanoBRET™ HTS Kinase Selectivity Assays

In NanoBRET™ assays, as the tracer and test compound bind to the target in a competitive and mutually exclusive manner, the relationship between the apparent IC50 of the test compound and the concentration of tracer used in the assay is governed by the Cheng–Prusoff equation (Eq. 1) [9]. For maximum quantitative accuracy, the assay is ideally run at a tracer concentration equal to or less than the apparent K_d ($K_{d,app}$) of the tracer, such that the measured IC_{50} of the test compound is within twofold of the apparent K_i ($K_{i,app}$), provided that this tracer concentration can render a sufficient assay window.

$$K_{i,app} = \frac{IC_{50}}{1 + \left(\frac{[tracer]}{K_{d,app}}\right)} \quad (1)$$

Transitioning assays from a 96-well to a 384-well format requires reoptimization of tracer concentration due to a possible shift in tracer performance and hence its $K_{d,app}$. To estimate tracer $K_{d,app}$, the IC50 of a test compound is measured and plotted at various tracer concentrations in a "matrix experiment." Subsequently, a linear regression is performed, and an algebraically transformed Cheng–Prusoff equation (Eq. 2) [10] is used to calculate $K_{d,app}$ of the tracer and $K_{i,app}$ of the test compound.

$$IC_{50} = K_{i,app} + \frac{K_{i,app}}{K_{d,app}}[tracer] \quad (2)$$

A sample plate layout for a matrix experiment is shown in Fig. 3. In this layout, four technical replicates of two different targets can easily fit on one 384-well assay plate, allowing calculation of assay window and Z′ at each tracer concentration [11] using Eqs. 3 and 4, respectively, using BRET values from DMSO-treated samples (DMSO) and samples treated with >10-fold excess of control compound (control).

$$\text{Assay window} = \frac{\text{Mean BRET}_{\text{DMSO}}}{\text{Mean BRET}_{\text{Contol}}} \quad (3)$$

$$Z' = 1 - \frac{3 * (\text{Stdev BRET}_{\text{DMSO}} + \text{Stdev BRET}_{\text{Control}})}{|\ \text{Mean BRET}_{\text{DMSO}} - \text{Mean BRET}_{\text{Contol}}\ |} \quad (4)$$

Table 1 lists the 192 kinase assays with their assay windows and Z′ at the given optimal tracer concentrations. Nearly all individual kinases yielded Z′ values well above the accepted cutoff for HTS ($Z' > 0.5$). This format proved successful for assays with both high and low assay windows (AW) as demonstrated in Fig. 4 for MAPK4 (AW = 13.6) and CHEK2 (AW = 1.9) (Fig. 4a, b). Both targets displayed increasing signal window with increasing tracer concentration, and the test compound, CC1 (a pan-kinase inhibitor), was capable of competing off the tracer and therefore decreasing the

Matrix Transfection Layout

	1	2	3	4	5	6	7	8	9	10	11	12	13	14	15	16	17	18	19	20	21	22	23	24
A	T1	T1	T1	T1	T1	T1	T1	T1	T1	T1	T1	T1	T1	T1	T1	T1	T1	T1	T1	T1	T1	T1	T1	T1
B	T1	T1	T1	T1	T1	T1	T1	T1	T1	T1	T1	T1	T1	T1	T1	T1	T1	T1	T1	T1	T1	T1	T1	T1
C	T2	T2	T2	T2	T2	T2	T2	T2	T2	T2	T2	T2	T2	T2	T2	T2	T2	T2	T2	T2	T2	T2	T2	T2
D	T2	T2	T2	T2	T2	T2	T2	T2	T2	T2	T2	T2	T2	T2	T2	T2	T2	T2	T2	T2	T2	T2	T2	T2
E	T1	T1	T1	T1	T1	T1	T1	T1	T1	T1	T1	T1	T1	T1	T1	T1	T1	T1	T1	T1	T1	T1	T1	T1
F	T1	T1	T1	T1	T1	T1	T1	T1	T1	T1	T1	T1	T1	T1	T1	T1	T1	T1	T1	T1	T1	T1	T1	T1
G	T2	T2	T2	T2	T2	T2	T2	T2	T2	T2	T2	T2	T2	T2	T2	T2	T2	T2	T2	T2	T2	T2	T2	T2
H	T2	T2	T2	T2	T2	T2	T2	T2	T2	T2	T2	T2	T2	T2	T2	T2	T2	T2	T2	T2	T2	T2	T2	T2
I	T1	T1	T1	T1	T1	T1	T1	T1	T1	T1	T1	T1	T1	T1	T1	T1	T1	T1	T1	T1	T1	T1	T1	T1
J	T1	T1	T1	T1	T1	T1	T1	T1	T1	T1	T1	T1	T1	T1	T1	T1	T1	T1	T1	T1	T1	T1	T1	T1
K	T2	T2	T2	T2	T2	T2	T2	T2	T2	T2	T2	T2	T2	T2	T2	T2	T2	T2	T2	T2	T2	T2	T2	T2
L	T2	T2	T2	T2	T2	T2	T2	T2	T2	T2	T2	T2	T2	T2	T2	T2	T2	T2	T2	T2	T2	T2	T2	T2
M	T1	T1	T1	T1	T1	T1	T1	T1	T1	T1	T1	T1	T1	T1	T1	T1	T1	T1	T1	T1	T1	T1	T1	T1
N	T1	T1	T1	T1	T1	T1	T1	T1	T1	T1	T1	T1	T1	T1	T1	T1	T1	T1	T1	T1	T1	T1	T1	T1
O	T2	T2	T2	T2	T2	T2	T2	T2	T2	T2	T2	T2	T2	T2	T2	T2	T2	T2	T2	T2	T2	T2	T2	T2
P	T2	T2	T2	T2	T2	T2	T2	T2	T2	T2	T2	T2	T2	T2	T2	T2	T2	T2	T2	T2	T2	T2	T2	T2

Matrix Tracer Concentration Layout

	1	2	3	4	5	6	7	8	9	10	11	12	13	14	15	16	17	18	19	20	21	22	23	24
A	C1	C1	C1	C1	C1	C1	C1	C1	C2	C2	C2	C2	C2	C2	C2	C2	C3	C3	C3	C3	C3	C3	C3	C3
B	C4	C4	C4	C4	C4	C4	C4	C4	C5	C5	C5	C5	C5	C5	C5	C5	C6	C6	C6	C6	C6	C6	C6	C6
C	C1	C1	C1	C1	C1	C1	C1	C1	C2	C2	C2	C2	C2	C2	C2	C2	C3	C3	C3	C3	C3	C3	C3	C3
D	C4	C4	C4	C4	C4	C4	C4	C4	C5	C5	C5	C5	C5	C5	C5	C5	C6	C6	C6	C6	C6	C6	C6	C6
E	C1	C1	C1	C1	C1	C1	C1	C1	C2	C2	C2	C2	C2	C2	C2	C2	C3	C3	C3	C3	C3	C3	C3	C3
F	C4	C4	C4	C4	C4	C4	C4	C4	C5	C5	C5	C5	C5	C5	C5	C5	C6	C6	C6	C6	C6	C6	C6	C6
G	C1	C1	C1	C1	C1	C1	C1	C1	C2	C2	C2	C2	C2	C2	C2	C2	C3	C3	C3	C3	C3	C3	C3	C3
H	C4	C4	C4	C4	C4	C4	C4	C4	C5	C5	C5	C5	C5	C5	C5	C5	C6	C6	C6	C6	C6	C6	C6	C6
I	C1	C1	C1	C1	C1	C1	C1	C1	C2	C2	C2	C2	C2	C2	C2	C2	C3	C3	C3	C3	C3	C3	C3	C3
J	C4	C4	C4	C4	C4	C4	C4	C4	C5	C5	C5	C5	C5	C5	C5	C5	C6	C6	C6	C6	C6	C6	C6	C6
K	C1	C1	C1	C1	C1	C1	C1	C1	C2	C2	C2	C2	C2	C2	C2	C2	C3	C3	C3	C3	C3	C3	C3	C3
L	C4	C4	C4	C4	C4	C4	C4	C4	C5	C5	C5	C5	C5	C5	C5	C5	C6	C6	C6	C6	C6	C6	C6	C6
M	C1	C1	C1	C1	C1	C1	C1	C1	C2	C2	C2	C2	C2	C2	C2	C2	C3	C3	C3	C3	C3	C3	C3	C3
N	C4	C4	C4	C4	C4	C4	C4	C4	C5	C5	C5	C5	C5	C5	C5	C5	C6	C6	C6	C6	C6	C6	C6	C6
O	C1	C1	C1	C1	C1	C1	C1	C1	C2	C2	C2	C2	C2	C2	C2	C2	C3	C3	C3	C3	C3	C3	C3	C3
P	C4	C4	C4	C4	C4	C4	C4	C4	C5	C5	C5	C5	C5	C5	C5	C5	C6	C6	C6	C6	C6	C6	C6	C6

Matrix Parent Compound Concentration Layout

	1	2	3	4	5	6	7	8	9	10	11	12	13	14	15	16	17	18	19	20	21	22	23	24
A	C1	C2	C3	C4	C5	C6	C7	D	C1	C2	C3	C4	C5	C6	C7	D	C1	C2	C3	C4	C5	C6	C7	D
B	C1	C2	C3	C4	C5	C6	C7	D	C1	C2	C3	C4	C5	C6	C7	D	C1	C2	C3	C4	C5	C6	C7	D
C	C1	C2	C3	C4	C5	C6	C7	D	C1	C2	C3	C4	C5	C6	C7	D	C1	C2	C3	C4	C5	C6	C7	D
D	C1	C2	C3	C4	C5	C6	C7	D	C1	C2	C3	C4	C5	C6	C7	D	C1	C2	C3	C4	C5	C6	C7	D
E	C1	C2	C3	C4	C5	C6	C7	D	C1	C2	C3	C4	C5	C6	C7	D	C1	C2	C3	C4	C5	C6	C7	D
F	C1	C2	C3	C4	C5	C6	C7	D	C1	C2	C3	C4	C5	C6	C7	D	C1	C2	C3	C4	C5	C6	C7	D
G	C1	C2	C3	C4	C5	C6	C7	D	C1	C2	C3	C4	C5	C6	C7	D	C1	C2	C3	C4	C5	C6	C7	D
H	C1	C2	C3	C4	C5	C6	C7	D	C1	C2	C3	C4	C5	C6	C7	D	C1	C2	C3	C4	C5	C6	C7	D
I	C1	C2	C3	C4	C5	C6	C7	D	C1	C2	C3	C4	C5	C6	C7	D	C1	C2	C3	C4	C5	C6	C7	D
J	C1	C2	C3	C4	C5	C6	C7	D	C1	C2	C3	C4	C5	C6	C7	D	C1	C2	C3	C4	C5	C6	C7	D
K	C1	C2	C3	C4	C5	C6	C7	D	C1	C2	C3	C4	C5	C6	C7	D	C1	C2	C3	C4	C5	C6	C7	D
L	C1	C2	C3	C4	C5	C6	C7	D	C1	C2	C3	C4	C5	C6	C7	D	C1	C2	C3	C4	C5	C6	C7	D
M	C1	C2	C3	C4	C5	C6	C7	D	C1	C2	C3	C4	C5	C6	C7	D	C1	C2	C3	C4	C5	C6	C7	D
N	C1	C2	C3	C4	C5	C6	C7	D	C1	C2	C3	C4	C5	C6	C7	D	C1	C2	C3	C4	C5	C6	C7	D
O	C1	C2	C3	C4	C5	C6	C7	D	C1	C2	C3	C4	C5	C6	C7	D	C1	C2	C3	C4	C5	C6	C7	D
P	C1	C2	C3	C4	C5	C6	C7	D	C1	C2	C3	C4	C5	C6	C7	D	C1	C2	C3	C4	C5	C6	C7	D

Fig. 3 Sample layout of matrix experiments. Four matrix experiments were run for two different protein targets on each 384-well assay plate with 6 concentrations of tracer and an 8-point concentration–response curve of the test compound. *T1/T2* target protein 1/target protein 2, *C1–C7* concentration 1 through concentration 7, *D* DMSO control

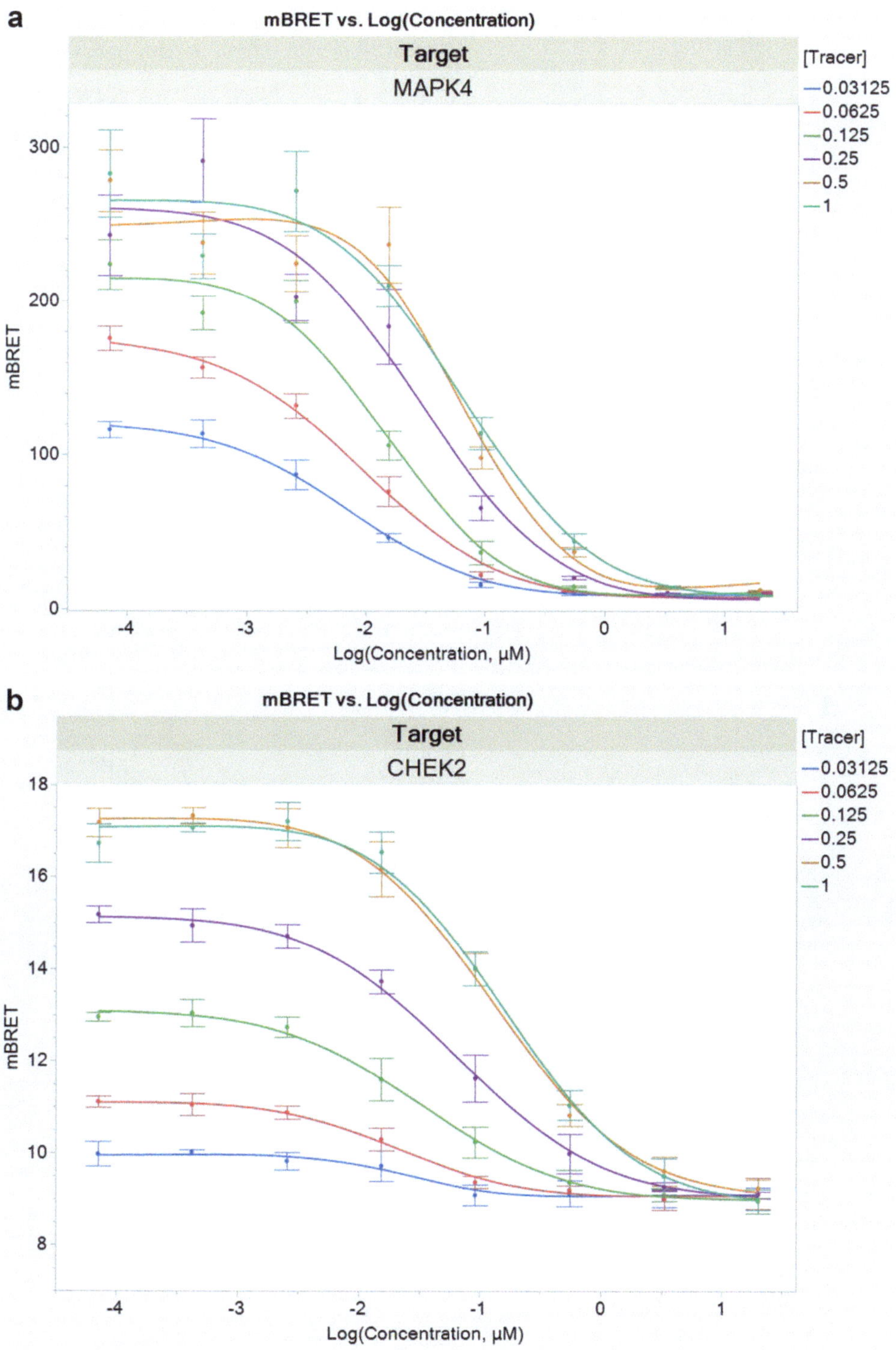

Fig. 4 Optimization of NanoBRET™ TE assays through Cheng-Prusoff analysis. (**a**, **b**) HEK293 cells were transiently transfected with MAPK4 and CHEK2 NanoLuc® fusion proteins respectively. Cells were treated with an 8-point, threefold dilution of CC1 parental compound using a 6-point, twofold dilution series of tracer K10 (shown as individual curves). (**c**, **d**) IC50 values were obtained from each curve within the matrix experiment and plotted vs. tracer concentrations for linear Cheng–Prusoff analysis for MAPK4 and CHEK2 respectively. (**e**) Table of resulting IC50 values of CC1 at each tested tracer concentration

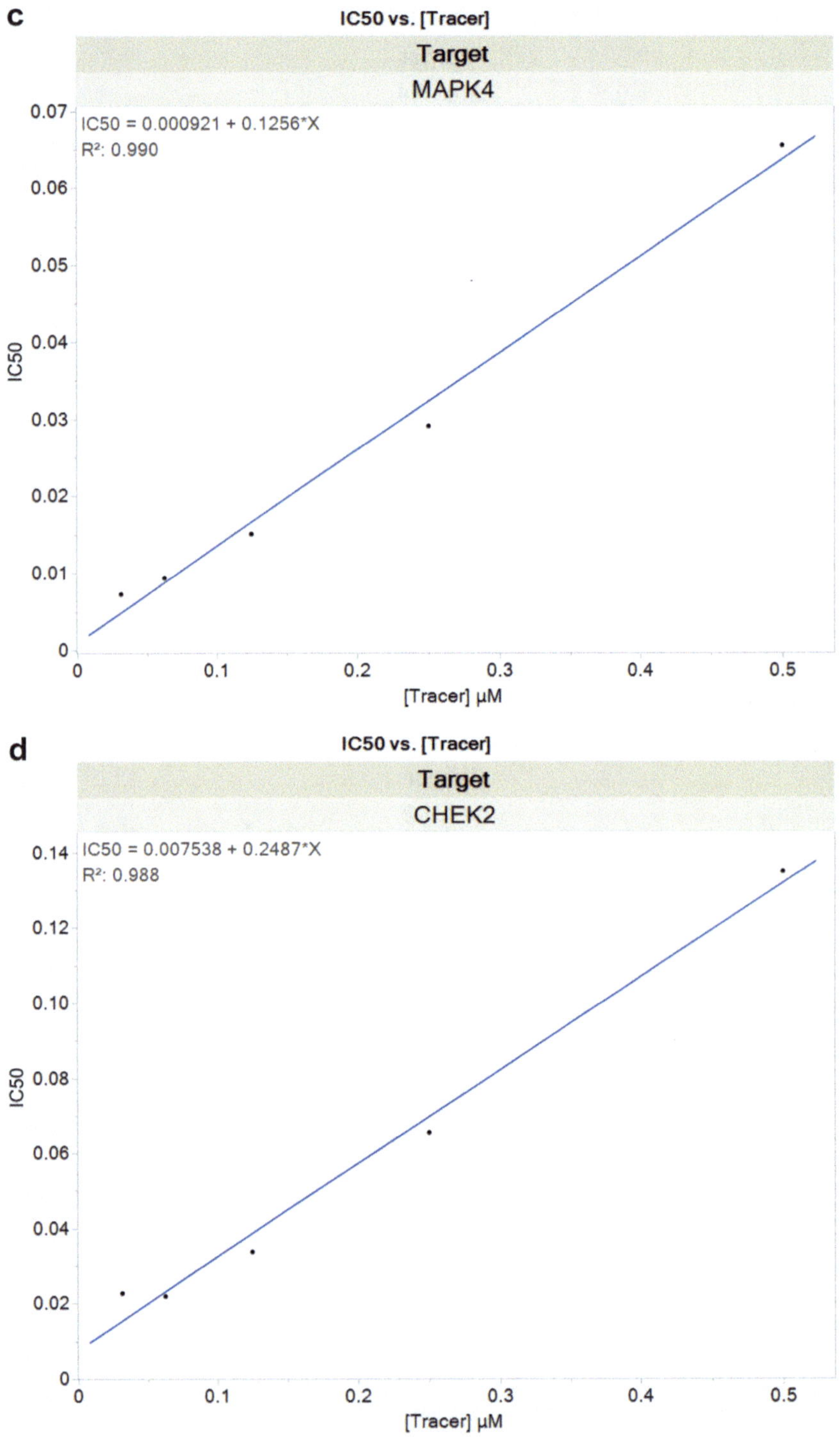

Fig. 4 (continued)

e

Tabulate

	Target	
	CHEK2	MAPK4
	IC50	IC50
Tracer Concentration (μM)	Mean	Mean
0.03125	0.02265	0.00726
0.0625	0.02185	0.00936
0.125	0.03369	0.01507
0.25	0.06549	0.02908
0.5	0.13490	0.06549

Fig. 4 (continued)

NanoBRET™ signal in a concentration-dependent manner (*See* **Note 3**). We were able to calculate $K_{d,app}$ and $K_{i,app}$ for both MAPK4 and CHEK2 (Fig. 4c–e) using the Cheng–Prusoff relationship described in Eq. 2.

The following protocol describes standard procedures used in our laboratory to perform NanoBRET™ TE assays in 384-well format.

1. Prepare compound and tracer dilutions in pure DMSO according to solubility limits at 400–1000× the final assay concentration.
   (a) A single source concentration can be used to generate different assay concentrations depending on the volume dispensed.
   (b) Additional dilutions may be necessary to achieve lower assay concentrations.
2. Dispense tracer and compounds into source plates qualified for acoustic dispensing. Low-volume plates can be used to conserve compounds.
3. Use an acoustic dispenser, such as the Beckman Coulter Echo, to dispense 20–50 nL of compound and tracer into assay plates containing transfected cells (Prepared in Subheading 3.2 Optimizing Cell Preparation and Transfection for HTS Kinase Selectivity Profiling).
4. Incubate assay plates at 37 °C and 5% CO_2 for 2 h prior to NanoBRET™ detection.
5. Prepare 3× NanoBRET™ Nano-Glo® Substrate with Extracellular NanoLuc® Inhibitor according to manufacturer guidelines.

6. Remove plates from the incubator for 10 min to acclimate to room temperature.
7. Dispense 10 μL of NanoBRET™ Nano-Glo® Substrate into each well using an automated dispenser, such as the BioTek MultiFlo FX.
8. Incubate the assay plate at room temperature for 2–3 min.
9. Measure donor emission (e.g., 450 nm) and acceptor emission (e.g., 610 nm or 630 nm) using a NanoBRET™-compatible luminometer.
10. Generate raw BRET ratio values by dividing the acceptor emission value by the donor emission value.
11. Convert raw BRET ratio values to milliBRET units (mBRET) by multiplying each raw BRET ratio value by 1000.

3.4 Validating Performance of Individual Kinase Assays in Measuring Compound Binding Affinity in Concentration–Response Experiments

A concentration–response assay is required for accurate measurement of the binding affinity of a compound to a target in NanoBRET™ TE assays. A good assay Z′ value (determined in Subheading 3.3) ensures reliable assay performance in single-dose HTS; however, Z′ cannot predict the reproducibility and variance of the assay in measuring compound binding affinity or potency in concentration–response assays [12]. Instead, the performance in concentration–response assays is evaluated using minimum significant ratio (MSR) [12, 13]. The MSR is a statistical parameter defined as the smallest ratio between the potencies of two compounds that is statistically significant. The MSR can be obtained experimentally using a control compound or a collection of test compounds [13].

It is noteworthy that each individual kinase assay within the NanoBRET™ TE K192 Kinase Selectivity System has unique performance characteristics and robustness (owing to variable assay windows, intrinsic noise factors, etc.). Therefore, following tracer concentration optimization and assay verification through Z′ analysis, we validated all 192 kinase assays with control compound MSR [13]. To obtain the control compound MSR, measurements of binding affinity of a control compound with known activity were collected in multiple biological replicates over time. The control compound MSR was then calculated using Eq. 5, where *S* was the standard deviation of the log potency of the control compound across the biological replicates.

$$\mathrm{MSR} = 10^{2\sqrt{2}S} \tag{5}$$

In this new scalable workflow, we have established control compound MSRs by testing the concentration response of a pan-kinase inhibitor CC1 in multiple biological replicates for each of 192 kinases assays in the NanoBRET™ TE K192 Kinase Selectivity System. Figure 5 shows a sample plate layout for one such

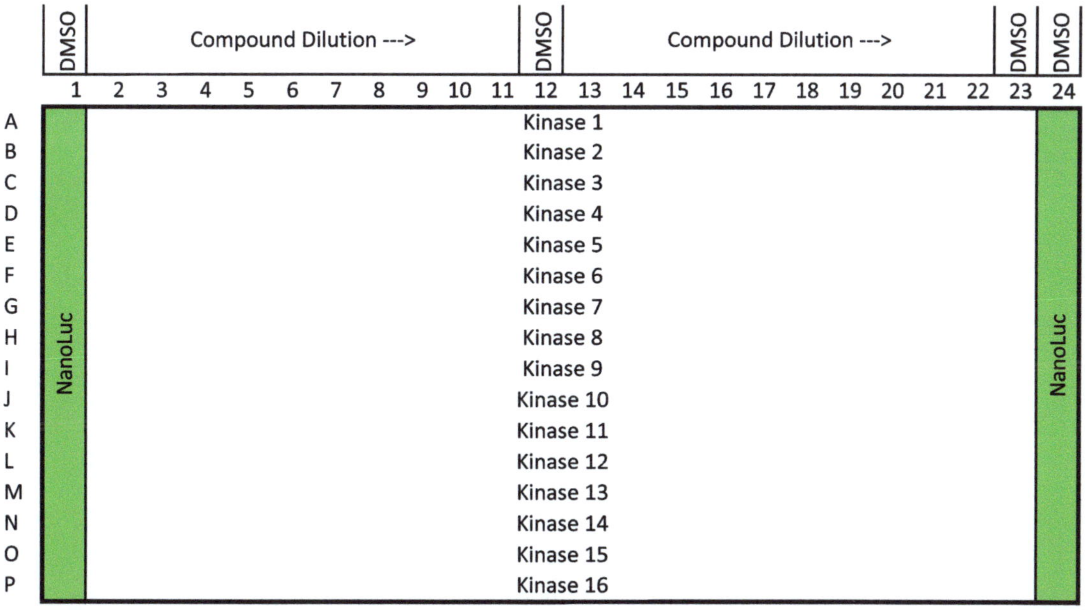

Fig. 5 Sample plate layout for concentration–response curve experiments

typical concentration–response experiment. Each plate has 16 kinase targets, and an 11-point concentration–response curve of CC1 was tested for each target in duplicate. Figure 6a and Table 1 show the distribution of MSR and the individual MSR values for the 192 kinase targets. 97% (187/192) of the kinase targets had control compound MSRs lower than the empirical acceptance value of 7.5 [13], and 75% were lower than 3.5, demonstrating that a majority of the assays are highly reproducible in estimating binding affinity, at least for the control compound. The control compound MSR experiment can also help identify assays that require significant improvement before being used for routine IC50 profiling. In Fig. 6a, a subset of targets (5/192) had control compound MSRs greater than 7.5 indicating they were less suitable as profiling assays using the test condition. Upon further investigation, it was commonly observed that, for most of the targets with high MSR, the concentration–response curves were lacking an upper or lower asymptote using the compound concentrations tested in this screen, which led to inaccurate estimations of the IC_{50} values. By adjusting the concentration range of the control compound or using a more potent control compound in the concentration–response experiment, the MSR of those assays should be improved. For example, CLK1 had the largest MSR in our initial analysis (MSR = 178120) because in the first two experiments, the CC1 control compound was not diluted to low enough concentrations to generate lower asymptotes (Fig. 6b). We modified the dilution scheme in subsequent experiments to improve 4PL curve fitting. After excluding the first two experiments from

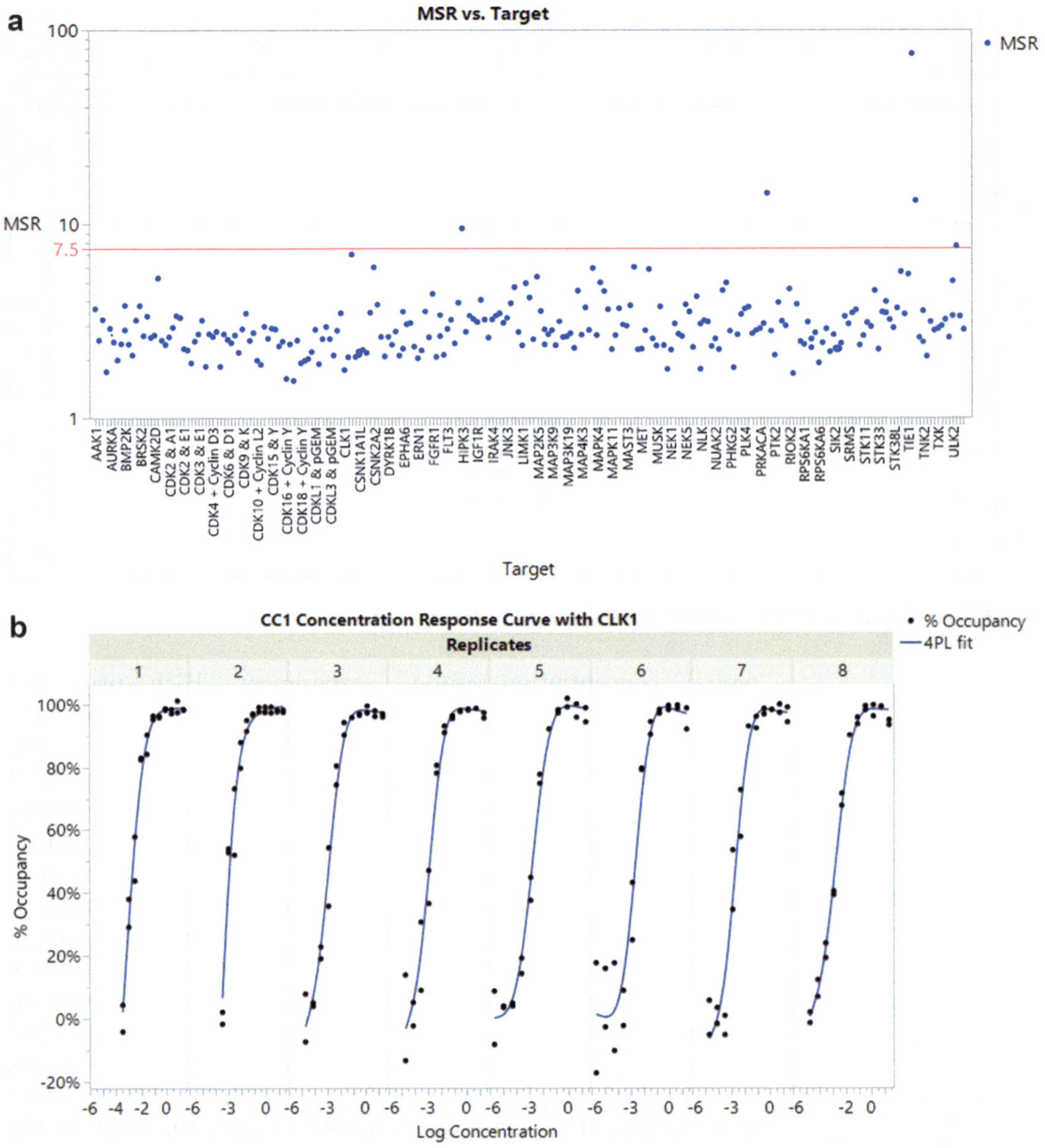

Fig. 6 MSR analysis to evaluate assay performance in concentration–response experiments. (**a**) Distribution of MSR values for the K192 kinase panel. (**b**) CC1 concentration response curves with CLK1 in multiple biological replicates. The curves in the first two replicates did not have lower asymptotes and contributed to high variance of LogIC50 estimates. Excluding data from these two experiments significantly improved MSR. (**c**) CC1 concentration response curves with TLK1 in multiple biological replicates. CC1 was not potent enough to generate upper asymptotes; therefore, the variance of LogIC50 was high. A more potent compound is required for generating complete concentration–response curves and improving MSR

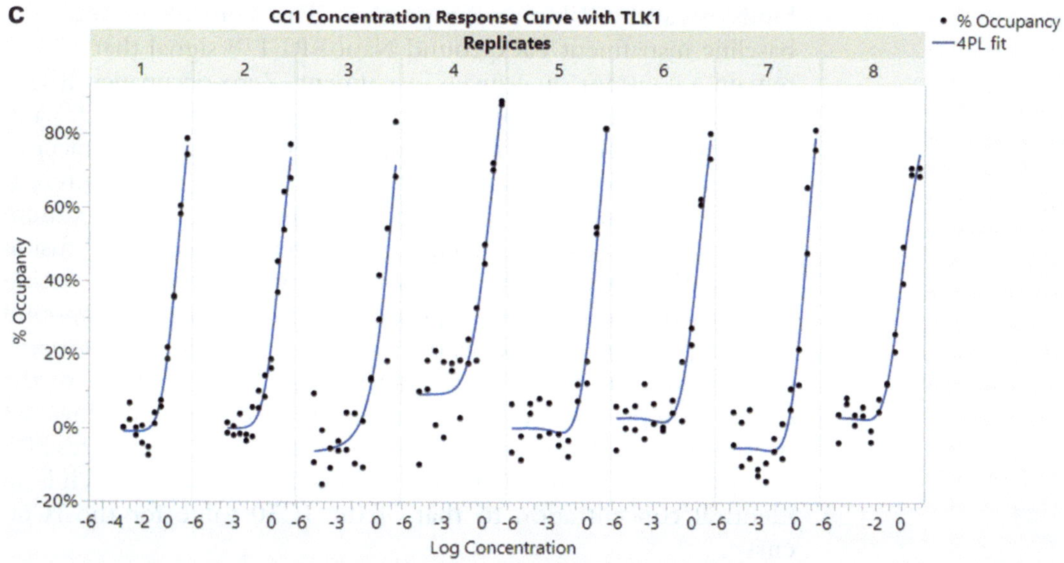

Fig. 6 (continued)

the analysis, the MSR was improved to 1.8. TLK1 is another example (Fig. 6c). CC1 was not potent enough to generate upper asymptotes in concentration–response experiments. A different control compound would be necessary to more accurately estimate MSR.

3.5 Evaluating Percent Occupancy of Compounds in Single-Dose Selectivity Profiling

To query intracellular selectivity of numerous test compounds against a broad cohort of kinase targets, single concentration profiling may be preferred over concentration–response analysis. In such an inhibitor selectivity profiling experiment, a test compound at a single concentration is surveyed against a large number of targets, and % occupancy of the compound for each target is then calculated using Eq. 6. The accuracy and variance of this measurement are dictated by the different performance of different kinase assays and can be evaluated using the data from the control compound MSR experiment (as shown in Subheading 3.4).

$$\%\text{Occupancy} = 100 * \left[1 - \frac{\text{mBRET}_{\text{sample}} - \text{mBRET}_{\text{full occupancy control}}}{\text{mBRET}_{\text{zero occupancy control}} - \text{mBRET}_{\text{full occupancy control}}}\right] \quad (6)$$

In each assay plate of the control MSR experiment (Fig. 5), there are built-in controls to calculate full occupancy and zero-occupancy BRET values. Full occupancy BRET values are calculated using cells transfected with NanoLuc® not fused to any target protein. This reflects a condition in which the compound of interest binds to all available target protein. This control establishes a baseline instrument background NanoBRET™ signal that should remain a constant on a given instrument. Zero-occupancy BRET values are calculated using cells treated with DMSO (vehicle) and tracer, which reflects the maximum possible binding of the tracer to target protein molecules in the absence of a compound of interest. A zero-occupancy control is necessary to gauge the maximum energy transfer possible through each tracer and NanoLuc® fusion combination, which provides an upper limit to the assay. Using Eq. 6, each concentration–response curve in the control compound MSR experiment can be mathematically converted to a % occupancy curve. From the % occupancy curve, concentrations of the control compound that give rise to specific % occupancy values can be interpolated. For example, the concentration that generates 50% occupancy will be the EC50 of the % occupancy curve, which is an identical concentration to that of the IC50 value for the IC50 curve.

To evaluate the variance of the assay in estimating % occupancy at a single concentration, we first calculated the mean LogEC50 and LogEC30 values of % occupancy curves from the multiple biological replicates in the control compound MSR experiment. Then, we used the mean LogEC50 and LogEC30 to interpolate % occupancy values at respective concentrations for each replicate from the corresponding 4PL regressions. Next, we performed statistical analysis to evaluate the variance of the interpolated % occupancy. Figure 7 shows the distribution (Fig. 7a) and CV% (Fig. 7b) of the interpolated % occupancy at mean LogEC50 and LogEC30 for the 192 kinase assays. The variance of the interpolated % occupancy values differs greatly among different targets, illustrating the different performances of different assays. At mean LogEC50, the average CV% of interpolated % occupancy for all 192 kinases was 19%, and 80% of them had CV% less than 23%, indicating 50% occupancy is a robust criterion for target engagement for most targets in this automated setting. At mean LogEC30, not surprisingly, most of the assays become noisier, and the average CV% of interpolated % occupancy was much higher at 38%. Very high CV% of an assay is often caused by poor fitting of an incomplete concentration–response curve and may not imply poor performance of the assay *per se*. To get a more accurate estimate of LogEC50 and thus the variance of the interpolated % occupancy, repeating the concentration–response experiment with a different concentration range or a different control compound is required. For assays with intrinsic high CV% or in the instance when low occupancy is observed, a follow-up concentration–response experiment would be necessary to better qualify target engagement. By analyzing the variance of the interpolated % occupancy,

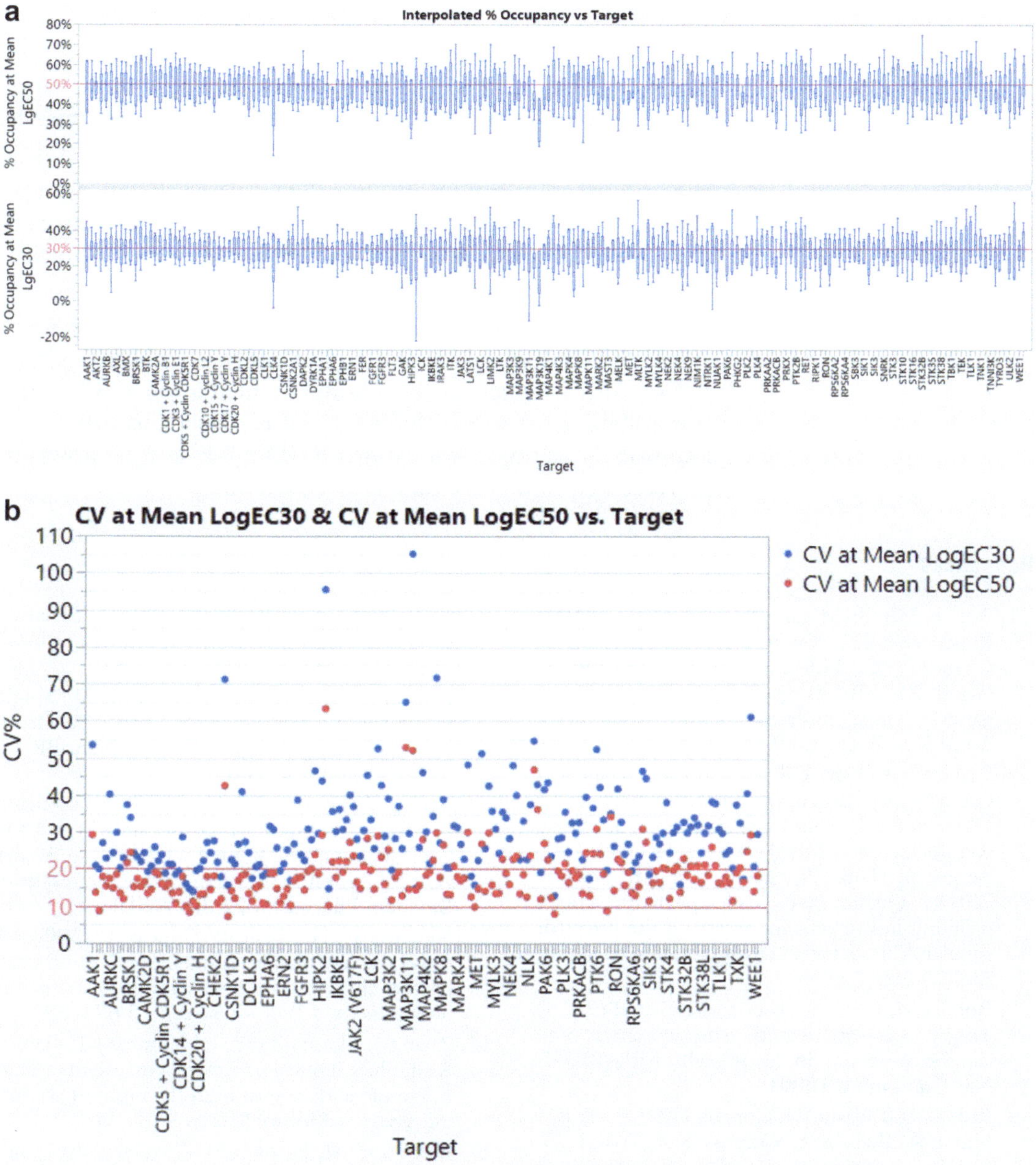

Fig. 7 Interpolated % occupancy at mean LogEC50 and mean LogEC30 of concentration–response curves. (**a**) Box and whisker plots showing the distribution of the interpolated % occupancy for each target. Top panel corresponds to 50% occupancy and bottom panel for 30% occupancy (red threshold line across the graph). (**b**) CV% of the interpolated % occupancy. The data used to generate these graphs are also listed in Table 1

the robustness and reproducibility of the assays in detecting hits in a single-dose profiling experiment can be evaluated. This workflow of statistical analysis can be performed in any laboratory setting equipped with similar automation.

4 Notes

1. Assay drift has been seen with thaw-and-use cells but is target specific. Test all new batches to ensure adequate performance.
2. Transfection reagent:DNA ratios should be reoptimized from the original assay conditions. For the kinase selectivity example shown in Fig. 2, a 3:1 Fugene® HD:DNA ratio was utilized in the 96-well format, but a 4:1 ratio was required for efficient transfection in the 384-well format.
3. It is critical to ensure that binding equilibrium between the tracer and compound at the target of interest has been achieved before BRET detection. Two hours of incubation at 37 °C in a humidified incubator with 5% CO_2 after tracer and compound addition is standard for NanoBRET™ TE assays; however, increased equilibration time may be needed.

References

1. Robers MB, Friedman-Ohana R, Huber KVM, Kilpatrick L, Vasta JD, Berger B-T, Chaudhry C, Hill S, Muller S, Knapp S, Wood KV (2020) Quantifying target occupancy of small molecules within living cells. Annu Rev Biochem 89:557–581
2. Vasta JD, Corona CR, Wilkinson J, Zimprich CA, Hartnett JR, Ingold MR, Zimmerman K, Machleidt T, Kirkland TA, Huwiler KG, Ohana RF, Slater M, Otto P, Cong M, Wells CI, Berger B, Hanke T, Glas C, Ding K, Drewry (2018) Quantitative, wide-spectrum kinase profiling in live cells for assessing the effect of cellular ATP on target engagement. Cell Chem Biol 25:206–214
3. Robers M, Dart M, Woodroofe C et al (2015) Target engagement and drug residence time can be observed in living cells with BRET. Nat Commun 6:10091
4. Robers MB, Vasta JD, Corona CR, Ohana RF, Hurst R, Jhala MA, Comess KM, Wood KV (2019) Quantitative, real-time measurements of intracellular target engagement using energy transfer. In: Methods in molecular biology, pp 45–71
5. Wells C, Vasta J et al (2020) Quantifying CDK inhibitor selectivity in live cells. Nat Commun. https://doi.org/10.1038/s41467-41020-16559-41460
6. Davies MI et al (2011) Comprehensive analysis of kinase inhibitor selectivity. Nat Biotechnol. https://doi.org/10.1038/nbt.1990
7. Karaman M et al (2008) A quantitative analysis of kinase inhibitor selectivity. Nat Biotechnol. https://doi.org/10.1038/nbt1358
8. Robers MB, Wilkinson JM et al (2021) Single tracer-based protocol for broad-spectrum kinase profiling in live cells with NanoBRET. STAR Ptotoc 2(4):100822
9. Cheng Y-C, Prusoff WH (1973) Relationship between the inhibition constant (KI) and the concentration of inhibitor which causes 50 per cent inhibition (I50) of an enzymatic reaction. Biochem Pharmacol 22:3099–3108
10. Newton P, Harrison P, Clulow S (2008) A novel method for determination of the affinity of protein: protein interactions in homogeneous assays. J Biomol Screen 13:674–682
11. Zhang J, Chung TDY, Oldenburg KR (1999) A simple statistical parameter for use in evaluation and validation of high throughput screening assays. J Biomol Screen 4:67–73
12. Eastwood BJ, Farmen MW, Iversen PW, Craft TJ, Smallwood JK, Garbison KE, Delapp NW, Smith GF (2006) The minimum significant ratio: a statistical parameter to characterize the reproducibility of potency estimates from concentration-response assays and estimation by replicate-experiment studies. J Biomol Screen 11:253–261
13. Haas JV, Eastwood BJ, Iversen PW, et al (2013) Minimum significant ratio – a statistic to assess assay variability. Assay Guidance Manual [Internet].

Chapter 9

A Fluorescence-Based Reporter Gene Assay to Characterize Nuclear Receptor Modulators

Espen Schallmayer and Daniel Merk

Abstract

Reporter gene assays are critical tools of nuclear receptor research for characterizing the effects of ligands on nuclear receptor activity. Common luciferase-based techniques require expensive substrates and are typically performed in endpoint format. Here, we describe a versatile reporter gene assay to observe nuclear receptor activity with fluorescent proteins as reporters. This setting is highly cost-efficient and enables observation of nuclear receptor activity over time with multiple measurements from one plate.

Key words Fluorescent proteins, Time-resolved activity, Cost-efficient, Luciferase

1 Introduction

The nuclear receptor (NR) family comprises 48 ligand-activated transcription factors divided into seven subfamilies [1, 2]. NRs bind to DNA as monomers, homodimers, or heterodimers and regulate gene expression in response to ligand binding. Thereby, NRs are involved in multiple processes such as embryogenesis, homeostasis, growth, and cell proliferation and differentiation [2]. Therefore, the protein family represents a class of attractive drug targets for various indications. Despite the therapeutic potential of NRs, only about half the NR family is well studied and covered by chemical tools and drugs. Knowledge on several so-called orphan receptors is limited, and ligands to study their roles are lacking [3].

Ligand discovery and development for NRs require robust and economic in vitro assay systems to observe nuclear receptor modulation by ligands. Two types of assay systems are commonly used. There are cell-free assays, which observe co-regulator binding or dimerization in a cell-free setting. These assays rely on labeled NR ligand binding domain protein and labeled co-regulators and determine NR modulation by changes in NR-co-regulator interaction observed by FRET or AlphaScreen. As cellular assays, reporter gene

Daniel Merk and Apirat Chaikuad (eds.), *Chemogenomics: Methods and Protocols*, Methods in Molecular Biology, vol. 2706,
https://doi.org/10.1007/978-1-0716-3397-7_9,

assays are common to characterize NR modulators. The cellular context provides the advantage that only compounds entering the cell can activate reporter gene transcription and that this important aspect of NR ligands can be observed in compound optimization [4].

Many reporter gene assays rely on luciferases as reporter genes, which offer strong and stable signal, but suffer from high costs for luciferase substrates. Additionally, most luciferase-based reporter gene assays cannot be easily employed to monitor NR activation over time. In this chapter, we present the protocol for a versatile cellular assay to determine NR activity using fluorescent proteins as reporter genes. By overcoming the need for luciferase substrates and cell lysis [5], this assay is highly cost-efficient and enables straightforward observation of NR activation or repression over time.

2 Materials

2.1 Plasmids

1. pFA-CMV-NR-LBD, encoding a Gal4 hybrid of the NR in question, based on pFA-CMV (Agilent Technologies 219036). For details on cloning, refer to [5].
2. Gal4 responsive mCherry reporter; e.g., pUAS-mCherry-NLS (Addgene #87695).
3. eGFP plasmid with constitutively active promoter for mammalian cells; e.g., SV40-eGFP-Z1 (Addgene #127638).

2.2 Cell Culture

1. HEK293T cells (e.g., from ATCC).
2. Routine cell culture medium; DMEM (500 mL) high glucose supplemented with 10% fetal calf serum (FCS), sodium pyruvate (1 mM), penicillin (100 U/mL), and streptomycin (100 μg/mL).
3. Transfection medium; Opti-MEM without supplements.
4. Assay medium; Opti-MEM supplemented with penicillin (100 U/mL) and streptomycin (100 μg/mL).
5. Trypsin–EDTA solution; 5 mg/mL trypsin and 2 mg/mL EDTA in 0.9% NaCl solution.
6. Trypan blue solution; 4 mg/mL trypan blue in 0.9% NaCl solution.
7. Invitrogen™ Lipofectamine™ LTX reagent with Plus™ reagent (Thermo Fisher Scientific, Cat. No.: 15338030).
8. Optional: coating solution; 0.1% collagen G in PBS.

2.3 Test Compounds and References

1. Reference agonist for the respective nuclear receptor as positive control (1000× stock in DMSO).
2. Test compounds (1000× stock of highest test concentration in DMSO).
3. DMSO.

2.4 Equipment

1. Fully equipped sterile work bench for cell culture and assay work.
2. Incubator at 37 °C and 5 Vol.% CO_2 for mammalian cells.
3. Cell counter or Neubauer chamber for cell counting at seeding.
4. Clear 96-well plates; e.g., clear CELLSTAR®, PS, sterile, F-bottom (Greiner Bio-One, Cat. No.: 655180).
5. Light microscope.
6. Plate reader for fluorescence measurement (e.g., Tecan Spark® Microplate Reader).
7. Single channel pipets (0.5–10 μL, 10–100 μL, 20–200 μL, 100–1000 μL).
8. Multichannel pipet (30–300 μL) or repetitive pipet with suitable tip to dispense 100 μL.
9. Stopwatch.

2.5 Software

1. Microsoft Excel.
2. Software for nonlinear regression (e.g., GraphPad Prism, Origin, Sigma Plot).

3 Methods

3.1 General Considerations

The following procedure refers to an assay optimized for 96-well format in HEK293T cells. Use of other formats and/or other cell lines will require suitable adaptions. The assay described here typically requires 3 days with an average of 2–3 h per day and allows to observe NR activation over time or in endpoint format. Figure 1 gives a schematic overview of the assay procedure.

3.2 Seeding Cells (0 h)

1. Grow HEK293T cells (e.g., in a 175-cm^2 flask) in routine cell culture medium at 37 °C and 5 Vol.% CO_2 to 70–80% confluence.
2. Optional: Coat plates by adding 100 μL coating solution to each well, incubate at 37 °C for 30 min, and remove coating solution right before seeding (*see* **Note 1**).
3. Aspirate cell culture medium, rinse with 5 mL PBS, and aspirate again.

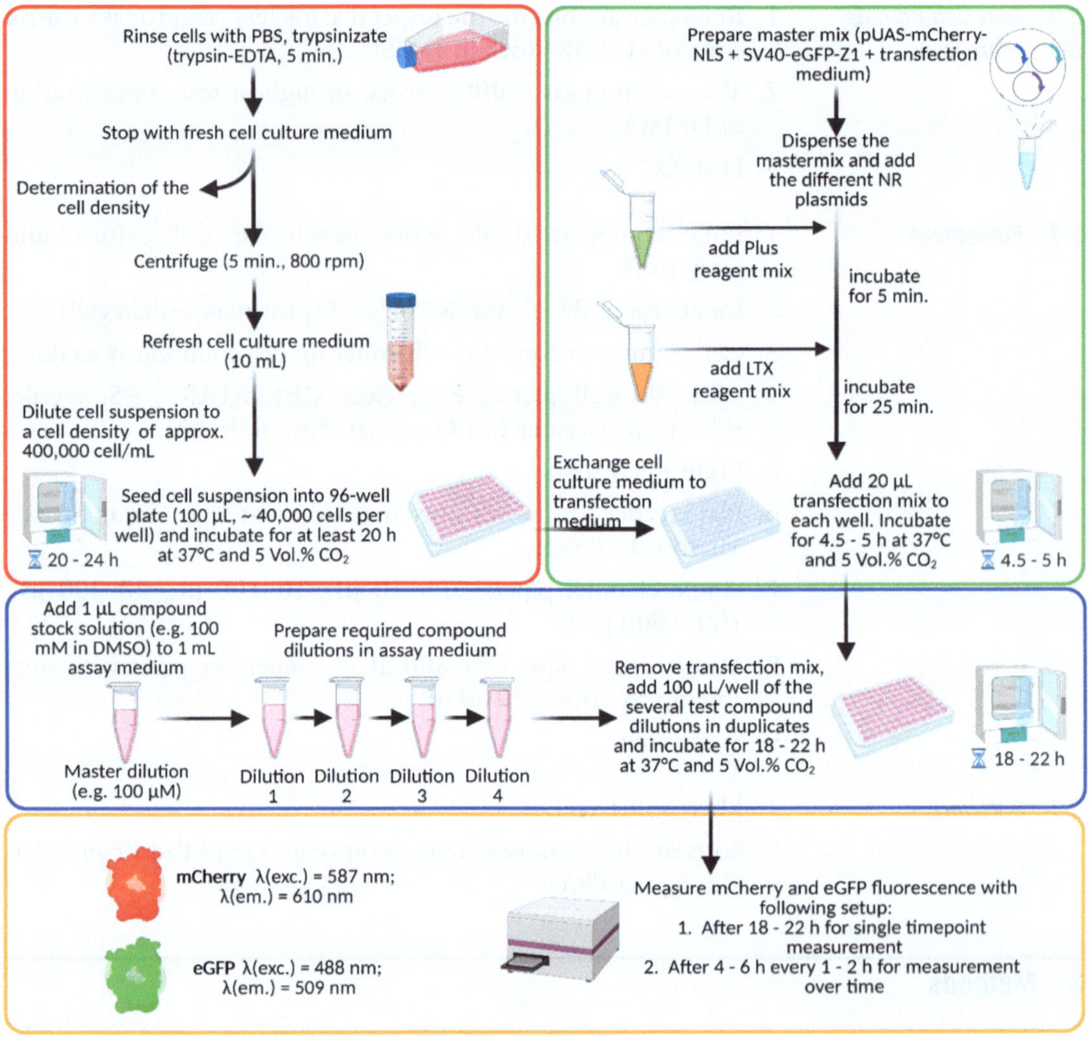

Fig. 1 Schematic overview of the assay procedure. Red box: preparation of the cells for cell seeding (see Subheading 3.2); green box: transfection procedure (see Subheading 3.3); blue box: representative procedure for test compound dilution and incubation (see Subheading 3.4); yellow box: measurement (see Subheading 3.5). (Created with BioRender.com)

4. Incubate with 5 mL of 0.5% Trypsin–EDTA for 5 min at 37 °C and 5 Vol.% CO_2.
5. Add 5 mL of fresh cell culture medium to stop and mix by pipetting up and down.
6. Incubate 10 µL of the cell suspension with 10 µL trypan blue solution for 1 min, add 10 µL of the suspension to counting chamber, and count the living cells.
7. Centrifuge remaining cell suspension (5 min, 800 rpm), remove supernatant medium, and resuspend cells in 10 mL of fresh cell culture medium.

Table 1
Exemplary plasmid amounts for a thyroid hormone receptor α (THRα) assay

Plasmid construct	Plasmid amount (ng/well)
pUAS-mCherry-NLS	100
SV40-eGFP-Z1	25
pFA-CMV-THRα-LBD	1

8. Dilute cell suspension to a cell density of approx. 400,000 cells/mL.
9. Seed cells in clear 96-well plates with 40,000 cells and 100 μL volume per well (cell density 400,000 cells/mL).
10. Keep cells in the incubator for at least 20 h at 37 °C and 5 Vol.% CO_2.

3.3 Transfection (24 h)

The following procedure refers to the use of multiple different pFA-CMV-NR-LBD clones in one assay. Therefore, a master mix with reporter and control plasmids is prepared and then split before the respective pFA-CMV-NR-LBD clones are added. Table 1 gives example plasmid amounts for a thyroid hormone receptor α (THR-α) [6] assay in 96-well format.

3.3.1 Preparations

1. Work under a sterile workbench.
2. Prepare plasmid master mix with appropriate amounts of pUAS-mCherry-NLS and SV40-eGFP-Z1 in transfection medium (15.1 μL/well). **Note 2** shows an Excel template for planning.
3. Split the master mix (15.1 μL/well) and add the respective pFA-CMV-NR-LBD (*see* **Note 2**).
4. Prepare Lipofectamine reagent mixtures (0.12 μL/well Plus™ reagent + 1.88 μL/well transfection medium; 0.20 μL/well LTX reagent + 2.70 μL/well transfection medium); exemplary template in **Note 3**.
5. Replace routine cell culture medium on 96-well plate with transfection medium (100 μL/well).

3.3.2 Transfection Procedure

Work under a sterile workbench. Repeat the following procedure for each receptor-specific plasmid mix.

1. Start transfection procedure by adding 2.0 μL/well Plus™ reagent mix to the plasmid mix.
2. Incubate for 5 min at room temperature, do not vortex.
3. Add 2.9 μL/well LTX reagent mix to the plasmid + Plus reagent mix.

4. Incubate 25 min at room temperature, do not vortex.
5. After the total incubation time (30 min), add 20 μL/well of transfection mixtures to each well.
6. Incubate cells for 4.5–5 h at 37 °C and 5 Vol.% CO_2.

3.4 Test Compound Dilution and Incubation

3.4.1 Test Compound Dilution (27 h)

1. All compound preparations and dilutions should be sterile; work under a sterile workbench.
2. Prepare stock solutions of test compounds in DMSO at 1000× concentration of the highest concentration to be tested (e.g., 100 mM).
3. Prepare master dilutions (e.g., 100 μM) of test compounds by adding 1 μL of the stock solution to 1 mL assay medium.
4. Prepare assay medium containing 0.1% DMSO for dilution series; use this medium also as negative control (at least one sample per plate) in the assay.
5. Prepare a dilution series for each test compound in assay medium containing 0.1% DMSO according to the intended concentration series (*see* **Note 4** for an example template).
6. For antagonistic testing, prepare assay medium containing a fixed concentration (e.g., 1 μM) of the reference agonist for the dilution series.
7. Prepare reference agonist solutions with appropriate concentration (e.g., 1 μM) in assay medium containing 0.1% DMSO; use the reference agonists as positive control (at least one sample per plate) in the assay.

3.4.2 Incubation with Test Compounds (29 h)

1. Work under a sterile workbench.
2. After 4.5–5 h of incubation with transfection mix, remove the transfection mix from the 96-well plate, and replace it with the appropriate dilutions of the test compounds (100 μL/well).
3. Do not remove medium from more than 24 wells at a time.
4. Test each compound concentration at least in duplicates; repeat in at least three independent experiments.
5. Include at least one negative (assay medium containing 0.1% DMSO) and one positive control sample (reference agonist for the receptor of interest at $\geq EC_{90}$, e.g., 1 μM) for each assay condition (i.e., receptor) and plate.
6. Incubate at 37 °C and 5 Vol.% CO_2 for 18–22 h for single time point (i.e., endpoint) measurements.
7. To observe NR modulation over time, measure fluorescence (*see* Subheading 3.5) every 1–2 h starting 4–6 h after incubation; keep plates at 37 °C and 5 Vol.% CO_2 between measurements.

Table 2
Exemplary parameters for mCherry and eGFP fluorescence measurements

eGFP measurement	
λ(exc.) = 488 nm	λ(em.) = 509 nm
λ(exc.) bandwidth = 5 nm	λ(em.) bandwidth = 5 nm
Number of flashes: 30; integration time: 40 μs; lag time: 0 μs; settle time: 25 ms	
mCherry measurement	
λ(exc.) = 587 nm	λ(em.) = 610 nm
λ(exc.) bandwidth = 5 nm	λ(em.) bandwidth = 5 nm
Number of flashes: 30; integration time: 40 μs; lag time: 0 μs; settle time: 25 ms	

Table 3
Data processing for assay raw data

Result type	Equation
Relative fluorescence units (RFUs)	(Fluorescence (mCherry))/(fluorescence (eGFP)) × 10,000
Fold activation	(RFU (test compound))/(RFU (negative control))
Relative activation	(Fold activation (test compound))/(fold activation (positive control)) × 100

3.5 Fluorescence Measurement (36–72 h)

1. Table 2 shows exemplary measurement parameters; different plate readers might require adaptions; optimize Z-position; and potentially gain at first measurement.
2. Determine mCherry fluorescence (λ(exc.) = 587 nm; λ(em.) = 610 nm).
3. Determine eGFP fluorescence (λ(exc.) = 388 nm; λ(em.) = 509 nm).

3.6 Data Analysis

3.6.1 Data Analysis for Individual Assays

Use Microsoft Excel or a similar software to calculate relative fluorescence units (RFU), fold activation, and relative activation for each well according to Table 3.

3.6.2 Data Analysis per Compound

To obtain EC_{50} or IC_{50} values, record dose–response curves and plot test compound concentrations against RFU, fold activation, or relative activation data and fit a sigmoidal curve by nonlinear regression (e.g., in GraphPad). A dose–response curve should cover about eight (at least six) concentrations over several orders of magnitude and include values in the baseline, the slope, and the upper plateau. Use data from at least three independent experiments. Either determine EC_{50} or IC_{50} for each experiment or use mean ± SD data for curve fitting. The dose–response curve provides EC_{50} or IC_{50} value and maximum (relative) activation or repression of the respective NR by the respective test compound.

4 Notes

1. HEK293T cells adhere weakly to cell culture plates, but a uniform monolayer is required for the best assay results. Coating plates with Collagen G before cell seeding may improve assay results.
2. Planning of transfection can be performed in Excel. Table 4 gives an exemplary Excel template for planning plasmid mixtures for an assay in 96-well format. The upper table shows Excel internal references, and the lower table gives an example with exemplary values. The table gives the respective amounts of transfection medium (cell H10), pUAS-mCherry-NLS (cell H2), and SV40-eGFP-Z1 (cell H3) needed to prepare the master mix (green cells) based on the total number of wells needed (cell F2–F3 and F10–F11). Cells K4–K8 show, how the master mix is split for the different conditions (different NR) based on the desired number of wells (F4–F8, blue cells). The respective pFA-CMV-NR-LBD plasmid amounts to be added to the previously split master mix are shown in cells G4–G8 (orange cells).

Table 4
Excel template for planning transfection

A	B	C	D	E	F	G	H	I	J	K
1			c(stock)/µg/µL	ng/well	wells	µL 1:10 stock	µL stock			
2		pUAS-mCherry-NLS	0,610	100	=SUM(F4:F8)	=E2*F2/D2/1000	=IF(G2>=0.5;G2; IF(G2=0; "-"; "x"))			Mastermix µL
3		SV40-eGFP-Z1	0,335	25	=SUM(F4:F8)	=E3*F3/D3/1000	=IF(G3>=0.5;G3; IF(G3=0; "-"; "x"))			
4		Receptor 1	0,014	1	30	=E4*F4/D4/1000	=IF(G4>=0.5;G4; IF(G4=0; "-"; "x"))		Receptor 1	=((F4/$F$11)*$H$11)
5		Receptor 2	0,068	6	4	=E5*F5/D5/1000	=IF(G5>=0.5;G5; IF(G5=0; "-"; "x"))		Receptor 2	=((F5/$F$11)*$H$11)
6		Receptor 3	0,123	12	10	=E6*F6/D6/1000	=IF(G6>=0.5;G6; IF(G6=0; "-"; "x"))		Receptor 3	=((F6/$F$11)*$H$11)
7		Receptor 4	0,231	25	21	=E7*F7/D7/1000	=IF(G7>=0.5;G7; IF(G7=0; "-"; "x"))		Receptor 4	=((F7/$F$11)*$H$11)
8		Receptor 5	0,312	50	32	=E8*F8/D8/1000	=IF(G8>=0.5;G8; IF(G8=0; "-"; "x"))		Receptor 5	=((F8/$F$11)*$H$11)
9										
10		Transfection medium		15.1 µL/well	=SUM(F4:F8)		=15.1*F10			
11		Mastermix			=SUM(F4:F8)		=SUM(H10;H2:H3)			

A	B	C	D	E	F	G	H	I	J	K
1			c(stock)/µg/µL	ng/well	wells	µL 1:10 stock	µL stock			
2		pUAS-mCherry-NLS	0,610	100	97	15,91	15,91			Mastermix µL
3		SV40-eGFP-Z1	0,335	25	97	7,25	7,25			
4		Receptor 1	0,014	1	30	2,19	2,19		Receptor 1	15,3
5		Receptor 2	0,068	6	4	0,35	x		Receptor 2	92,0
6		Receptor 3	0,123	12	10	0,98	0,98		Receptor 3	184,1
7		Receptor 4	0,231	25	21	2,27	2,27		Receptor 4	383,5
8		Receptor 5	0,312	50	32	5,13	5,13		Receptor 5	766,9
9										
10		Transfection medium		15.1 µL/well	97		1.464,70			
11		Mastermix			97		1.487,86			

The upper table shows the internal Excel references, and the lower table shows an example. Cells referring to the master mix are green, cells showing how the master mix is split for different conditions are blue, and cells showing the receptor plasmid amounts to be added for different conditions are orange

Table 5
Excel template for planning transfection

A	B	C	D	E	F	G	H	I	J	K
12	PLUS			µL/well	wells		µL			PLUS µL
13	regent	PLUS		0.12	=F10		=E13*F13		Receptor 1	=F4*2
14	mix	Transfection medium		1.88	=F10		=E14*F14		Receptor 2	=F5*2
15									Receptor 3	=F6*2
16									Receptor 4	=F7*2
17									Receptor 5	=F8*2
18										
19	LTX			µL/well	wells		µL			LTX µL
20	reagent	LTX		0.2	=F10		=E20*F20		Receptor 1	=F4*2.9
21	mix	Transfection medium		2.7	=F10		=E21*F21		Receptor 2	=F5*2.10
22									Receptor 3	=F6*2.11
23									Receptor 4	=F7*2.12
24									Receptor 5	=F8*2.13

A	B	C	D	E	F	G	H	I	J	K
12	PLUS			µL/well	wells		µL			PLUS µl
13	regent	PLUS		0.12	97		11.64		Receptor 1	60
14	mix	Transfection medium		1.88	97		182.36		Receptor 2	8
15									Receptor 3	20
16									Receptor 4	42
17									Receptor 5	64
18										
19	LTX			µL/well	wells		µL			LTX µl
20	reagent	LTX		0.2	97		19.40		Receptor 1	87
21	mix	Transfection medium		2.7	97		261.90		Receptor 2	11.6
22									Receptor 3	29
23									Receptor 4	60.9
24									Receptor 5	92.8

The upper table shows the internal Excel references, and the lower table shows an example. Column H contains the volumes of medium and reagent for the reagent mixes, and column K shows the amounts of each reagent mix to be added to each plasmid mix

3. Table 5 gives an exemplary Excel template for planning transfection reagent mixtures (Plus™ and LTX reagent mixtures) for an assay in 96-well format. The upper table shows Excel internal references, and the lower table gives an example with exemplary values. To prepare the Plus™ reagent mixture, transfection medium (H14) is mixed with Plus™ reagent (H13) according to the total number of wells (F13, F14). The LTX reagent mixture is prepared in an analogous manner by adding LTX reagent (H20) to transfection medium (H21) according to the total number of wells (F20, F21). The volume of Plus™ reagent mixture to be added to the respective plasmid mixes is calculated in cells K13–K17. The volume of LTX reagent mixture to be added to the respective plasmid mixes is calculated in cells K20–K24.

4. Compound dilution series can be planned in Excel. Table 6 gives an exemplary Excel template for planning compound dilution. Starting point is a compound stock solution in DMSO. The first dilution step is the preparation of a master dilution (e.g., test compound 1) by pipetting 1 µL of stock solution to 1 mL of assay medium. Further dilutions are prepared by pipetting the respective volume from previous dilutions into assay medium containing 0.1% DMSO. For antagonistic testing (green), the assay medium contains a fixed concentration of a reference agonist (e.g., 1 µM).

Table 6
Excel template for planning test compound dilution series

Negative control				
Concentration [µM]	Stock solution [µL]	Assay medium [mL]	from previous [µL]	Assay medium + 0.1% DMSO [µL]
0	—	—	—	4000
Reference agonist (Stock solution = 10 mM)				
Concentration [µM]	Stock solution [µL]	Assay medium [mL]	from previous [µL]	Assay medium + 0.1% DMSO [µL]
10	1	1	—	—
1			100	900
Test compound 1 (Stock solution = 100 mM)				
Concentration [µM]	Stock solution [µL]	Assay medium [mL]	from previous [µL]	Assay medium + 0.1% DMSO [µL]
100	1	1	—	—
30			300	700
10			300	600
3			300	700
1			300	600
Test compound 2 (Stock solution = 100 mM)				
Concentration [µM]	Stock solution [µL]	Assay medium [mL]	from previous [µL]	Assay medium + 0.1% DMSO [µL]
100	1	1	—	—
50			500	500
30		from 100 µM	300	700
10			300	600
3			300	700
Test compound 3 (antagonist) (Stock solution = 100 mM)				
Concentration [µM]	Stock solution [µL]	Assay medium + reference agonist (1 µM) [mL]	from previous [µL]	Assay medium + reference agonist (1 µM) [µL]
100	1	1	—	—
10			100	900
1			100	900
0,3			300	700
0,1			300	600

References

1. Germain P, Staels B, Dacquet C, Spedding M, Laudet V (2006) Overview of nomenclature of nuclear receptors. Pharmacol Rev 58:685–704. https://doi.org/10.1124/pr.58.4.2
2. Aranda A, Pascual A (2001) Nuclear hormone receptors and gene expression. Physiol Rev 81: 1269–1304
3. Benoit G, Cooney A, Giguere V, Ingraham H, Lazar M, Muscat G, Perlmann T, Renaud JP, Schwabe J, Sladek F, Tsai MJ, Laudet V (2006) International union of pharmacology. LXVI. Orphan nuclear receptors. Pharmacol Rev 58: 798–836. https://doi.org/10.1124/pr.58.4.10
4. Merk D, Steinhilber D, Schubert-Zsilavecz M (2014) Characterizing ligands for farnesoid X receptor-available in vitro test systems for farnesoid X receptor modulator development. Expert Opin Drug Discov 9:27–37. https://doi.org/10.1517/17460441.2014.860129
5. Heering J, Merk D (1966) Hybrid reporter gene assays: versatile in vitro tools to characterize nuclear receptor modulators. Methods Mol Biol 2019:175–192. https://doi.org/10.1007/978-1-4939-9195-2_14
6. Gellrich L, Heitel P, Heering J, Kilu W, Pollinger J, Goebel T, Kahnt A, Arifi S, Pogoda W, Paulke A, Steinhilber D, Proschak E, Wurglics M, Schubert-Zsilavecz M, Chaikuad A, Knapp S, Bischoff I, Fürst R, Merk D (2020) L-thyroxin and the nonclassical thyroid hormone TETRAC are potent activators of PPARÎ. J Med Chem 63:6727–6740. https://doi.org/10.1021/acs.jmedchem.9b02150

Chapter 10

Measuring Protein–Protein Interactions in Cells using Nanoluciferase Bioluminescence Resonance Energy Transfer (NanoBRET) Assay

Magdalena M. Szewczyk, Dominic D. G. Owens, and Dalia Barsyte-Lovejoy

Abstract

Protein–protein interactions (PPIs) are increasingly recognized for their roles in functional cellular networks and their importance in disease-targeting contexts. Assessing PPI in the native cellular environment is challenging and requires specific and quantitative methods. Bioluminescence resonance energy transfer (BRET) is a biophysical process that can be used to quantify PPI. With Nanoluciferase bioluminescent protein as a donor and a fluorescent chloroalkane ligand covalently bound to HaloTag protein as an acceptor, NanoBRET provides a versatile and robust system to quantitatively measure PPI in living cells. BRET efficiency is proportional to the distance between the donor and acceptor, allowing for the measurement of PPI in real time. In this paper, we describe the use of NanoBRET to study specific interactions between proteins of interest in living cells that can be perturbed by using small-molecule antagonists and genetic mutations. Here, we provide a detailed protocol for expressing NanoLuc and HaloTag fusion proteins in cell culture and the necessary optimization of NanoBRET assay conditions. Our example results demonstrate the reliability and sensitivity of NanoBRET for measuring interactions between proteins, protein domains, and short peptides and quantitating the PPI antagonist compound activity in living cells.

Key words NanoBRET, Protein-protein interaction, Cellular assay, Protein-protein interaction antagonists' screen

1 Introduction

Protein–protein interactions (PPIs) are critical for most cellular pathways and processes, including signal transduction, gene regulation, and protein synthesis [1]. These interactions allow proteins to coordinate the diversity of cellular functions and are also involved in many diseases, including cancer [2]. In fact, several drugs currently in use are designed to disrupt or modulate specific PPIs in order to alter the course of a disease [3, 4]. Therefore,

Daniel Merk and Apirat Chaikuad (eds.), *Chemogenomics: Methods and Protocols*, Methods in Molecular Biology, vol. 2706,
https://doi.org/10.1007/978-1-0716-3397-7_10,

assessing PPI in cells is a key step toward understanding disease pathology and assessing cellular activity during drug development campaigns.

BRET (bioluminescence resonance energy transfer) is a biophysical process that can be used to assess PPI in living cells [5, 6]. It is based on the principle of resonance energy transfer, which occurs when a donor molecule transfers energy to an acceptor molecule through a non-radiative process [7]. In BRET, a donor protein that emits light (such as luciferase) is fused to one protein of interest, while an acceptor molecule that absorbs light (such as green fluorescent protein) is fused to another protein of interest [8]. If the two proteins of interest interact, energy from the donor is transferred to the acceptor, causing it to emit light. PPI can be measured by quantifying the light emitted by the acceptor molecule, as energy transfer reflects the interaction between donor and acceptor molecules [5, 7]. BRET can be used to study PPIs in a variety of cellular contexts and can provide important insights into the role of these interactions in biological processes and diseases [5].

NanoBRET is a variant of the BRET technique that is widely used to measure PPI in the cellular context [9–12]. In this configuration, one of the proteins of interest is tagged with NanoLuc (NL), a highly sensitive and bright 19.1 kDa bioluminescent protein from the deep-sea shrimp *Oplophorus gracilirostris* [13], which acts as the energy donor for the BRET reaction in the presence of the furimazine substrate (Fig. 1). The other protein of interest is tagged with a 33 kDa protein tag called HaloTag (HT) that binds covalently to chloroalkane-containing molecules [14]. A cell-permeable fluorescent chloroalkane molecule (HaloTag ligand) is added to cell culture media that binds to the HT protein and acts as the BRET energy acceptor. Like traditional BRET methods, the efficiency of the energy transfer is dependent on the distance

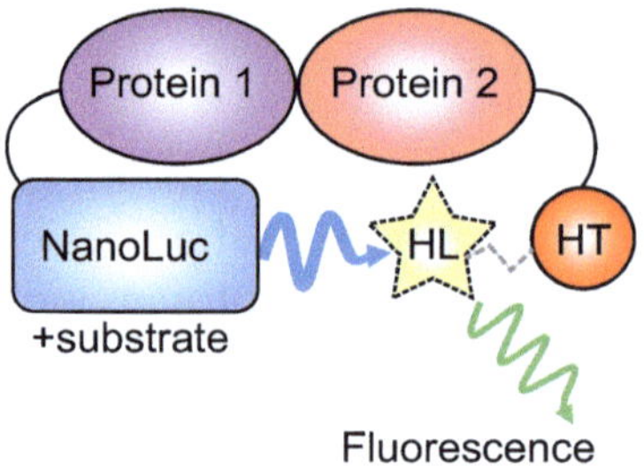

Fig. 1 Schematic of NanoBRET assay components. The interaction between hypothetical proteins 1 and 2 is to be measured by NanoBRET. Protein 1 is tagged with NanoLuc while protein 2 is tagged with HaloTag (HT). Fluorescent chloroalkane 618 Ligand (HL) covalently binds to HT. When protein 1 and protein 2 are in close proximity, in the presence of furimazine substrate, fluorescence is produced

between the two tagged proteins of interest, with Förster distances (R0) of 6–7 mm and a working range of 3–10 nm [15]. Thus, NanoBRET can be used to measure PPI in real time in living cells.

NanoBRET offers several advantages over other methods for detecting PPI. Unlike fluorescence resonance energy transfer (FRET) approaches, NanoBRET does not require an external excitation source and avoids issues surrounding phototoxicity and bleaching, direct excitation of the acceptor, and autofluorescence background [5]. Moreover, donor and acceptor signals are quantified independently in NanoBRET, acting as an internal control and accounting for discrepancies in protein expression levels across samples. This increases the reliability of NanoBRET compared with split luciferase complementation assays, where data are normalized to a separate reference sample, which can introduce additional sources of variability [16]. Also, the bright signal produced by NL allows quantification of PPI using proteins expressed at low, close to physiological levels [17]. This may be advantageous over other less bright bioluminescent donors that require higher expression levels, as artifacts may be observed when proteins are overexpressed at supraphysiological levels [18]. NanoBRET is applicable to high-throughput screening as it is compatible with both 96- and 384-well plate formats. As such, drug discovery is a particularly well-suited application of NanoBRET as PPIs can be quantified in the presence or absence of small-molecule antagonists to demonstrate cellular target engagement and cellular activity [19–21]. Overall, NanoBRET is a powerful tool for understanding the role of PPIs in biological processes and for developing new therapies for diseases.

2 Materials

2.1 Constructs

1. NanoBRET plasmid, either with N- or C-terminal NL or HT tag, generated by NanoBRET™ PPI MCS Starter System or NanoBRET™ PPI Flexi® Starter System (Promega).
2. Control constructs expressing HT protein and NL protein alone, generated by introducing stop codons in HT and NL N-terminal fusion vectors.

2.2 Specific Reagents

1. HaloTag® NanoBRET™ 618 Ligand (Promega).
2. Nano-Glo® Substrate (Promega).
3. Cell transfection reagent—any brand suitable for the cell type used.
4. Opti-MEM™ I Reduced Serum Medium (Gibco™).
5. Dulbecco's modified Eagle's medium (DMEM).
6. DMEM without phenol red.

7. Fetal bovine serum (FBS).
8. Penicillin and streptomycin.
9. White tissue culture (TC) plates (96-well or 384-well).
10. 6-well and 10-cm tissue culture plates.
11. 96-well PCR plate.
12. 0.25% Trypsin/EDTA.
13. PBS.

2.3 Instrumentation

A luminescence/fluorescence plate reader is capable of sequentially measuring dual-wavelength windows. Optimal filters are as follows: donor signal—band pass (BP) filter centered around 460 nm (e.g., emission 450 nm/BP80), acceptor signal—long pass (LP) filter starting around 600–610 nm (e.g., emission 610/LP). An example is a CLARIOstar microplate reader equipped with filters for donor 450 nm/80 nm BP and acceptor 610 nm LP (Mandel).

3 Methods

3.1 NanoBRET Optimization

3.1.1 Cell Plating

1. Culture HEK293T cells (*see* **Note 1**) in a 10 cm TC plate in DMEM, supplemented with 10% FBS, 100 U/mL penicillin, and 100 μg/mL streptomycin, up to 90% density. Overgrowing cells often result in an increased number of dead cells.
2. Remove the media from the plate and wash it with PBS. Add 1 mL of trypsin/EDTA, wait about 1 min, remove excess trypsin, and tap the dish to dislodge cells from the plastic. Resuspend cells in 1- to 2-mL culture media, and mix gently to generate a single-cell suspension. Count cell density using a hemocytometer or other cell counting technique and resuspend cells to a final density of 4×10^5 cells/mL (*see* **Note 2**) in the cell culture medium. Plate 100 μL cells per well in 96-well white TC plates and allow to attach and recover for 4–6 h at 37 °C, 5% CO_2 (*see* **Note 3**).

3.1.2 Cell Transfection

Initial optimization of NanoBRET is of critical importance. NanoBRET conditions must be empirically determined for each pair of proteins to determine the optimal tagging orientation (N- or C-terminal fusion proteins), which does not affect protein function and PPI and results in the highest BRET, as well as constructs ratio and expression levels. Avoid transfecting cells with high amounts of DNA, since high protein abundances may lead to aggregate formation inside cells, resulting in a nonspecific NanoBRET signal and high background. Use constructs expressing HT or/and NL protein alone to determine NanoBRET assay background levels. NanoBRET starter vector systems contain NanoBRET control pair, which may be used to test your instrumentation, following

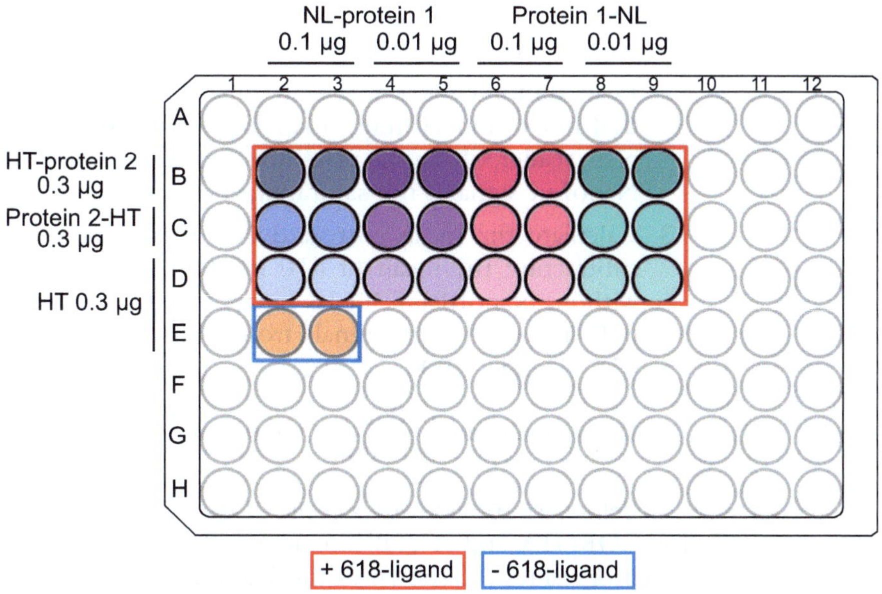

Fig. 2 Initial NanoBRET Optimization—μg DNA per 100 μL of Opti-MEM™ I Reduced Serum Medium. The example shows starting concentrations and tagging orientations (i.e., HT-tagged protein 2 is N-terminally tagged protein 2 and protein 2-HT is C-terminally tagged protein 2) for initial testing of NL-tagged protein 1 interaction with HT-tagged protein 2. Although this format is sufficient for most NanoBRET assays, further optimization of DNA concentrations, as well as alternate tagging (NL-tagged protein 2 and HT-tagged protein 1), may be required. Instead of HT alone control, NL alone controls can also be used

the manufacturer's instructions. As an initial starting point, we typically use 0.3 μg of HT vector with two different concentrations of NL vectors (0.1 and 0.01 μg, Fig. 2). The carrier DNA should be added to each transfection mix to reach a final concentration of 1 μg DNA per 100 μL of transfection mix (*see* **Notes 4** and **5**).

1. Prepare the transfection mixture according to the manufacturer's instructions (*see* **Note 6**). Warm up the reagent to room temperature (RT) before use. Resuspend 1 μg DNA in 100 μL Opti-MEM™ I Reduced Serum Medium, vortex briefly, add 2 μL of X-tremeGENE™ HP DNA Transfection Reagent (Roche) straight into solution, vortex for 10 s, and incubate for 15 min at RT. Add 10 μL of transfection mix per each of the wells in the 96-well, mix gently by shaking the plate, and keep at 37 °C, 5% CO_2 overnight. Transfect one or two additional wells with any NL and HT transfection mixture, which will be used for determining background level.

3.1.3 NanoBRET Measurement

1. The following day replace media with 40 μL of DMEM (without phenol red) supplemented with 4% FBS, 100 U/mL penicillin, and 100 μg/mL streptomycin and 1 μL/mL HaloTag® NanoBRET™ 618 Ligand. Add media without 618 Ligand to 1–2 control wells for background signal measurement. 2–4 h

later, add 10 μL/well of NanoBRET™ Nano-Glo® Substrate diluted 100-fold in DMEM with no phenol red (*see* **Note 7**).

2. Read donor emission at 450 nm and acceptor emission at 618 nm within 10 min of substrate addition using a suitable plate reader. Shake the plate 30 s before reading to ensure the NanoLuc signal is not saturated, and adjust gain accordingly.
3. Calculate the mean corrected NanoBRET ratios (mBU) by subtracting the mean of 610 nm/460 nm signal from cells without NanoBRET™ 618 Ligand × 1000 from the mean of 610 nm/460 nm signal from cells with NanoBRET™ 618 Ligand × 1000.

$$\mathrm{mBU} = \text{plus ligand}\frac{610\ \mathrm{nm}}{460\ \mathrm{nm}} * 1000 - \text{no ligand}\frac{610\ \mathrm{nm}}{460\ \mathrm{nm}} * 1000$$

Below, we provide an example of the optimization experiment for CORO1A and ACTB interaction. In Fig. 3, raw donor and acceptor values are shown as the NanoBRET ratios for indicated tagging configurations. All configurations gave NanoBRET ratio significantly higher over HT alone control; however, the best results were obtained between 0.03 μg/96-well N- and C-terminally HT-tagged ACTB and 0.001 μg/96-well C-terminally NL-tagged CORO1A (Fig. 3).

3.2 Testing PPI Antagonists in NanoBRET

Once the optimization is completed, the conditions giving the highest NanoBRET ratio as compared to HT or NL control can be used to test PPI antagonists. The compound treatment can be done in 96- or 384-well formats. The 384-well format is mostly used for higher throughput screening, e.g., testing several antagonists.

1. Prepare compounds and NanoBRET™ 618 Ligand solutions in DMEM (without phenol red) supplemented with 4% FBS, 100 U/mL penicillin, and 100 μg/mL streptomycin. Prepare serial dilutions of compounds in solvent, e.g., DMSO, so that working solutions of compounds maintain the same solvent concentration across all samples (*see* **Note 8**).
2. Culture HEK293T cells in a 10 cm TC plate in DMEM, supplemented with 10% FBS, 100 U/mL penicillin, and 100 μg/mL streptomycin, up to 90% density. Remove the media from the plate and wash it with PBS. Trypsinize and resuspend cells in culture media, and mix gently to generate a single-cell suspension. Resuspend cells to a final density of 4×10^5 cells/mL (*see* **Note 2**) in the cell culture medium. For 96-well format, plate 100 μL cells per well in 96-well white TC plates and for 384-well format 2 mL cells per well in transparent 6-well TC plates. Allow cells to attach and recover for 4–6 h at 37 °C, 5% CO_2.

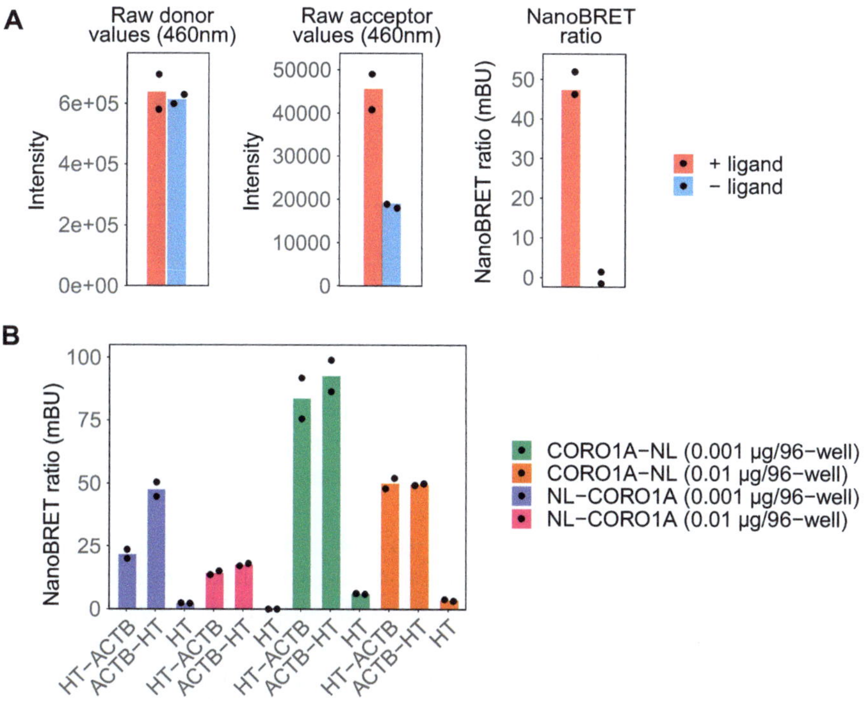

Fig. 3 NanoBRET assay optimization for CORO1A interaction with ACTB. (**a**) Example of NanoBRET ratio and raw donor and acceptor measurements with the NL-tagged CORO1A and HT-tagged ACTB. Raw donor values were measured using a 460 nm/80BP filter, and raw acceptor values were measured with a 610 nm/LP filter. Raw acceptor values from samples without NanoBRET™ 618 Ligand represent bleed-through and should be lower compared to samples containing ligand. The mean corrected NanoBRET ratios (mBU) are calculated by subtracting the mean of 610 nm/460 nm signal from cells without ligand × 1000 from the mean of 610/460 signal from cells with ligand × 1000. (**b**) Optimization of tag position and construct amounts. HEK293T cells were transfected with N- or C-terminally HT-tagged ACTB or HT alone (0.03 µg/96-well) and N- or C-terminally NL-tagged CORO1A (0.01 or 0.001 µg/96-well). The highest NanoBRET ratio is observed between C-terminally NL-tagged CORO1A (0.001 µg/96-well) and C-terminally HT-tagged ACTB

3. *96-Well Format*: The following day, prepare compound and diluent control working solutions at 2× the final desired concentration. Prepare 2 µL/mL (2×) NanoBRET™ 618 Ligand solution diluted in DMEM without phenol red. Add 20 µL of 2× NanoBRET™ 618 Ligand and 20 µL of 2× compound and diluent control working solutions to each well. Add media without 618 Ligand to 1–2 control wells that will provide the background measurements. Incubate for 4 h (37 °C, 5% CO_2) and then add 10 µL/well of NanoBRET™ Nano-Glo® Substrate diluted 100-fold in DMEM with no phenol red. Follow with **step 2** in Subheading 3.1.3.
4. *384-Well Format*: The following day trypsinize cells and resuspend in DMEM (10% FBS, 100 U/mL penicillin, and 100 µg/mL streptomycin). Pellet cells at 300× *g* for 2 min, remove media, and resuspend in DMEM (without phenol red)

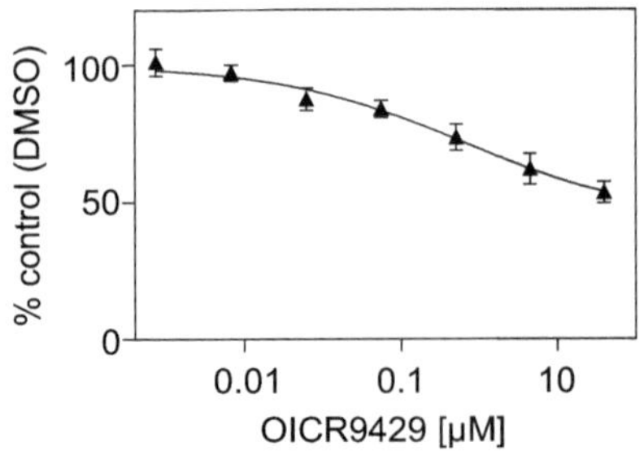

Fig. 4 Disruption of WDR5 interaction with histone H3 by OICR9429 compound. HEK293T cells were co-transfected with N-terminally NL-tagged WDR5 and C-terminally HT-tagged histone H3 for 20 h and treated with indicated concentrations of OICR9429 for 4 h. The graph represents the non-linear fit of mean corrected NanoBRET ratios of samples treated with compound normalized to DMSO control. The results are mean ± SD, $n = 4$, IC50 = 0.4 μM

supplemented with 4% FBS, 100 U/mL penicillin, and 100 μg/mL streptomycin to a final density of 2×10^5 cells/mL. Save 100 μL cell suspension for 2–4 wells without 618 Ligand control that will provide the background measurements. To the remaining cells, add 1 μL/mL HaloTag® NanoBRET™ 618 Ligand. Prepare 20× compound and diluent control working solutions. Transfer 95 μL/well cell suspension into a 96-well PCR plate and add 5 μL of 20× compound solutions. Mix gently and transfer 20 μL/well to 384-well plates in technical quadruplicates using a multichannel pipette. After incubating for 4 h, add 5 μL/well of NanoBRET™ Nano-Glo® Substrate diluted 100-fold in DMEM with no phenol red. Follow with **step 2** in Subheading 3.1.3.

Below, we provide an example of the dose response of the OICR9429 compound on disrupting the interaction between WD repeat protein 5 (WDR5) and histone H3 (Fig. 4). In the experiment, HEK293T cells were co-transfected with N-terminally NL-tagged WDR5 and C-terminally HT-tagged histone H3 for 20 h and treated with indicated concentrations of OICR9429 for 4 h. The dose-dependent decrease in the NanoBRET ratio was observed with increasing compound concentration with IC50 = 0.4 μM.

3.3 NanoBRET PPI Assays Using Isolated Domains and Short Peptide Sequences

NanoBRET can be utilized to measure interactions between full-length proteins and short peptides or isolated domains. To ensure interaction specificity and validate a positive result, the NanoBRET signal should be compared for the two proteins of interest and HT protein and/or NL protein alone.

The NanoBRET method can also be adapted not only for full-length proteins but for specific domains or even short peptide-based interactions (Figs. 5 and 6). It was previously published that DCAF1 interacts with VPR via the WD40 domain

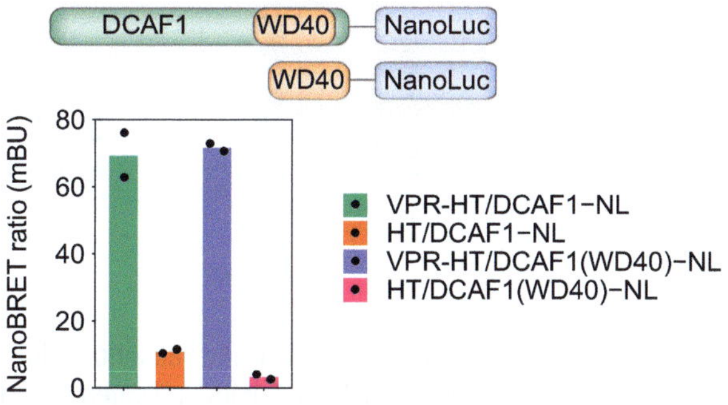

Fig. 5 NanoBRET assay with full-length and specific domain fusions. HEK293T cells were co-transfected with HT-tagged VPR and NL-tagged full-length DCAF1 or WD40 domain of DCAF1 (1038-1400 aa) for 20 h

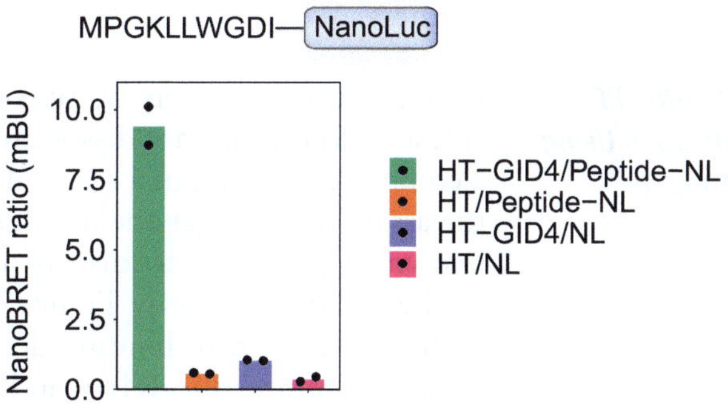

Fig. 6 NanoBRET assay with peptide fusion. HEK293T cells were co-transfected with HT-tagged GID4 and NL-tagged DDX50 N-terminal peptide (degron) for 20 h

[22]. When tested by NanoBRET, VPR interacted with both full-length DCAF1 and its WD40 domain alone, indicated by a higher NanoBRET signal compared to HT control (Fig. 5). PPI occurring through specific domains are usually easier to disrupt than interactions of full-length proteins that, by nature, may employ multiple surfaces. Domain-based NanoBRET usually results in lower background and higher NaoBRET assay windows. Thus, isolated protein domains can be used in PPI NanoBRET for examining the potency of domain-specific antagonists. Moreover, if a protein of interest is known to interact with short linear peptide sequences, peptide fusions may also be used to test PPI. For example, E3 ligases recruit substrates for proteasomal degradation by recognizing degrons—small, short amino acid sequences located anywhere in the protein [23]. Figure 6 shows GID4 (recognition component of CTLH E3 ligase [24, 25]) interacting with N-terminal peptide (degron) of DDX50 [26]. NanoBRET signal between the HT-GID4 and peptide-NL is significantly higher compared to the HT protein, or NL protein controls, demonstrating interaction

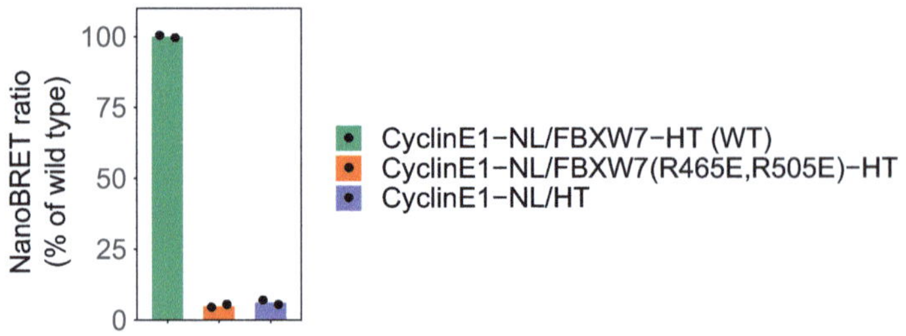

Fig. 7 NanoBRET assay validation with binding deficient mutants. HEK293T cells were transfected with indicated constructs for 20 h. The R465 and R505E mutation in FBXW7 protein abolished interaction with cyclin E1 (assay window 100%)

specificity. This format can be adapted for screening different peptides to determine binding specificity and finding potential binding partners.

3.4 NanoBRET PPI Assay Validation Using Genetic Mutants

Proteins usually exist in multiprotein complexes, and a small-molecule antagonist may not dissociate a protein of interest entirely from a complex, resulting in a residual NanoBRET signal. To determine the assay window, genetic mutations in donor or acceptor proteins known or predicted to disrupt the interaction can be utilized to further validate the assay. Figure 7 shows an example of NanoBRET assay validation with binding deficient mutants. Based on FBXW7 crystal structure, the R465 and R505 were predicted to play an important role in FBXW7 binding to its substrates [27]. The R465E and R505E double mutation in FBXW7 completely abolished its interaction with cyclin E1 indicated by the NanoBRET ratio being similar to WT FBXW7 interaction with unrelated HT protein (Fig. 7b).

4 Notes

1. Any cell type that may be readily transfected or transduced with the vectors for NL and HT can be used.
2. Cells larger than HEK293T, e.g., U2Os or HeLa should be plated at a lower density (about 2×10^5 cells/mL).
3. If not using a whole plate, avoid external wells, which are more prone to evaporation or temperature fluctuations, and fill them with 100 μL of PBS or media.
4. Use carrier DNA (e.g., empty pcDNA3.1 or pUC19 vector) to equalize the amount of total DNA transfected.
5. Inducible stable or stable expression of fusion proteins can also be considered.

6. Any transfection reagents suitable for chosen cell type can be used following the manufacturer's instructions.
7. Start the addition of Nano-Glo® solution from wells with no 618 Ligand to prevent cross-contamination since even a trace amount of 618 Ligand carried over will increase background signal levels.
8. Compound solvents such as DMSO may alter NanoBRET characteristics in a concentration-dependent manner. We recommend not exceeding 0.1% DMSO concentration and include control samples if DMSO concentrations are variable across the compound dilution series.

Acknowledgements

The Structural Genomics Consortium is a registered charity (no: 1097737) that receives funds from Bayer AG, Boehringer Ingelheim, Bristol Myers Squibb, Genentech, Genome Canada through Ontario Genomics Institute [OGI-196], EU/EFPIA/OICR/McGill/KTH/Diamond Innovative Medicines Initiative 2 Joint Undertaking [EUbOPEN grant 875510], Janssen, Merck KGaA (aka EMD in Canada and US), Pfizer, and Takeda.

References

1. Nooren IMA, Thornton JM (2003) Diversity of protein-protein interactions. EMBO J 22(14):3486–3492. https://doi.org/10.1093/emboj/cdg359
2. Arkin M (2005) Protein-protein interactions and cancer: small molecules going in for the kill. Curr Opin Chem Biol 9(3):317–324. https://doi.org/10.1016/j.cbpa.2005.03.001
3. Jin L, Wang W, Fang G (2014) Targeting protein-protein interaction by small molecules. Ann Rev Pharmacol Toxicol 54:435–456. https://doi.org/10.1146/annurev-pharmtox-011613-140028
4. Li B, Rong D, Wang Y (2019) Targeting protein-protein interaction with covalent small-molecule inhibitors. Curr Top Med Chem 19(21):1872–1876. https://doi.org/10.2174/1568026619666191011163410
5. Pfleger KDG, Eidne KA (2006) Illuminating insights into protein-protein interactions using bioluminescence resonance energy transfer (BRET). Nat Methods 3(3):165–174. https://doi.org/10.1038/nmeth841
6. Coriano C, Powell E, Xu W (2016) Monitoring ligand-activated protein-protein interactions using bioluminescent resonance energy transfer (BRET) assay. Method Mol Biol (Clifton, NJ) 1473:3–15. https://doi.org/10.1007/978-1-4939-6346-1_1
7. Wu P, Brand L (1994) Resonance energy transfer: methods and applications. Anal Biochem 218(1):1–13. https://doi.org/10.1006/abio.1994.1134
8. Boute N, Jockers R, Issad T (2002) The use of resonance energy transfer in high-throughput screening: BRET versus FRET. Trends Pharmacol Sci 23(8):351–354. https://doi.org/10.1016/s0165-6147(02)02062-x
9. Machleidt T, Woodroofe CC, Schwinn MK et al (2015) NanoBRET – a novel BRET platform for the analysis of protein-protein interactions. ACS Chem Biol 10(8):1797–1804. https://doi.org/10.1021/acschembio.5b00143
10. Dale NC, Johnstone EKM, White CW, Pfleger KDG (2019) NanoBRET: the bright future of proximity-based assays. Front Bioeng Biotechnol 7:56. https://doi.org/10.3389/fbioe.2019.00056
11. Groß VE, Gershkovich MM, Schöneberg T et al (2022) NanoBRET in C. elegans

illuminates functional receptor interactions in real time. BMC Mol Cell Biol 23(1):8. https://doi.org/10.1186/s12860-022-00405-w
12. Phillipou AN, Lay CS, Carver CE et al (2020) Cellular target engagement approaches to monitor epigenetic reader domain interactions. SLAS Discov Adv Life Sci R & D 25(2): 163–175. https://doi.org/10.1177/2472555219896278
13. Hall MP, Unch J, Binkowski BF et al (2012) Engineered luciferase reporter from a deep-sea shrimp utilizing a novel imidazopyrazinone substrate. ACS Chem Biol 7(11):1848–1857. https://doi.org/10.1021/cb3002478
14. Los GV, Encell LP, McDougall M et al (2008) HaloTag: a novel protein labeling technology for cell imaging and protein analysis. ACS Chem Biol 3(6):373–382. https://doi.org/10.1021/cb800025k
15. Weihs F, Wang J, Pfleger KDG, Dacres H (2020) Experimental determination of the bioluminescence resonance energy transfer (BRET) Förster distances of NanoBRET and red-shifted BRET pairs. Analytica Chimica Acta: X 6:100059. https://doi.org/10.1016/j.acax.2020.100059
16. Azad T, Tashakor A, Hosseinkhani S (2014) Split-luciferase complementary assay: applications, recent developments, and future perspectives. Anal Bioanal Chem 406(23): 5541–5560. https://doi.org/10.1007/s00216-014-7980-8
17. Mo X-L, Luo Y, Ivanov AA et al (2016) Enabling systematic interrogation of protein-protein interactions in live cells with a versatile ultra-high-throughput biosensor platform. J Mol Cell Biol 8(3):271–281. https://doi.org/10.1093/jmcb/mjv064
18. Sampaio NG, Kocan M, Schofield L et al (2018) Investigation of interactions between TLR2, MyD88 and TIRAP by bioluminescence resonance energy transfer is hampered by artefacts of protein overexpression. PLoS One 13(8):e0202408. https://doi.org/10.1371/journal.pone.0202408
19. Bradley WD, Arora S, Busby J et al (2014) EZH2 inhibitor efficacy in non-Hodgkin's lymphoma does not require suppression of H3K27 monomethylation. Chem Biol 21(11):1463–1475. https://doi.org/10.1016/j.chembiol.2014.09.017
20. Szewczyk MM, Ishikawa Y, Organ S et al (2020) Pharmacological inhibition of PRMT7 links arginine monomethylation to the cellular stress response. Nat Commun 11(1):2396. https://doi.org/10.1038/s41467-020-16271-z
21. Dilworth D, Hanley RP, Ferreira de Freitas R et al (2022) A chemical probe targeting the PWWP domain alters NSD2 nucleolar localization. Nat Chem Biol 18(1):56–63. https://doi.org/10.1038/s41589-021-00898-0
22. Wu Y, Zhou X, Barnes CO et al (2016) The DDB1-DCAF1-Vpr-UNG2 crystal structure reveals how HIV-1 Vpr steers human UNG2 toward destruction. Nat Struct Mol Biol 23(10):933–940. https://doi.org/10.1038/nsmb.3284
23. Varshavsky A (2019) N-degron and C-degron pathways of protein degradation. Proc Natl Acad Sci 116(2):358–366. https://doi.org/10.1073/pnas.1816596116
24. Liu H, Pfirrmann T (2019) The Gid-complex: an emerging player in the ubiquitin ligase league. Biol Chem 400(11):1429–1441. https://doi.org/10.1515/hsz-2019-0139
25. Maitland MER, Lajoie GA, Shaw GS, Schild-Poulter C (2022) Structural and functional insights into GID/CTLH E3 ligase complexes. Int J Mol Sci 23(11). https://doi.org/10.3390/ijms23115863
26. Dong C, Zhang H, Li L et al (2018) Molecular basis of GID4-mediated recognition of degrons for the Pro/N-end rule pathway. Nat Chem Biol 14(5):466–473. https://doi.org/10.1038/s41589-018-0036-1
27. Hao B, Oehlmann S, Sowa ME et al (2007) Structure of a Fbw7-Skp1-cyclin E complex: multisite-phosphorylated substrate recognition by SCF ubiquitin ligases. Mol Cell 26(1): 131–143. https://doi.org/10.1016/j.molcel.2007.02.022

Chapter 11

HiBiT Cellular Thermal Shift Assay (HiBiT CETSA)

Sarath Ramachandran, Magdalena Szewczyk, Samir H. Barghout, Alessio Ciulli, Dalia Barsyte-Lovejoy, and Victoria Vu

Abstract

Cellular thermal shift assay (CETSA) is based on the thermal stabilization of the protein target by a compound binding. Thus, CETSA can be used to measure a compound's cellular target engagement and permeability. HiBiT CETSA method is quantitative and has higher throughput compared to the traditional Western-based CETSA. Here, we describe the protocol for a HiBiT CETSA, which utilizes a HiBiT tag derived from the NanoLuciferase (NanoLuc) that upon complementation by LgBiT NanoLuc tag produces a bright signal enabling tracking of the effects of increasing temperature on the stability of a protein-of-interest in the presence/absence of various compounds. Exposure of a HiBiT-tagged protein to increasing temperatures induces protein denaturation and thus decreased LgBiT complementation and NanoLuc signal. As the stability of proteins at higher temperatures can be influenced by the compound binding, this method enables screening for target engagement in living or permeabilized cells.

Key words CETSA, HiBiT, Target engagement, Thermal shift, Split nanoluciferase, Drug discovery

1 Introduction

An important step in the drug discovery process is confirming the target engagement of a small molecule, thus ensuring binding to select target protein-of-interest (POI) in intact cells [1]. The quantification of compound binding to POI allows for determination of whether the biochemical or biophysical compound activity is correlated to cell activity and can aid in optimizing the structure–activity relationship to improve compound potency.

Cellular thermal shift assay (CETSA) is a powerful target engagement assay based on the principle that a protein's thermostability can be modified by its binding to a small molecule [2]. As a protein is exposed to increasing temperatures, it begins to denature and forms protein aggregates. However, if the protein binds a compound, it is stabilized and can withstand higher temperatures

Daniel Merk and Apirat Chaikuad (eds.), *Chemogenomics: Methods and Protocols*, Methods in Molecular Biology, vol. 2706, https://doi.org/10.1007/978-1-0716-3397-7_11,

before denaturing. This means the protein remains soluble and produces a thermal shift compared to in the absence of an interacting small molecule [2].

Traditionally, CETSA uses centrifugation to separate protein aggregates from soluble protein, followed by SDS-PAGE and Western blotting to quantify how much protein remains soluble at varying temperatures between compound-treated and untreated samples [3]. Recently, the NanoLuc Thermal Shift Assay (NaLTSA) was developed as a simpler and higher throughput way to screen for compound target engagement in cells than the traditional Western-based CETSA. The NaLTSA is based on a NanoLuc and POI fusion protein. The smaller NanoLuc (19 kDa) is a brighter and more stable luminescent protein with a longer half-life and greater activity than Firefly or Renilla luciferase [4]. By fusing NanoLuc with a POI, luminescence signal can be used to rapidly quantify remaining stable protein levels at varying temperatures. However, a caveat of NaLTSA is that the NanoLuc is still large enough that depending on the size of the POI, it can alter thermostability or introduce steric hindrance and affect resulting melt curves. Thus, the use of NanoLuc tagging needs to be thoroughly validated to determine if it affects target engagement and thermostability or introduces steric hindrance [5].

In the HiBiT CETSA, also known as the SplitLuc CETSA [6, 7] or BiTSA [8], a small tag is used in place of the modified NanoLuc protein by splitting the luciferase into an inactive luciferase subunit, Large Bit (LgBiT), and a small 11 amino acid HiBiT tag suitable for protein fragment complementation-based assays. A POI is fused with the HiBiT tag, and when a stable and soluble HiBiT-POI fusion is reconstituted with LgBiT, they form a functional NanoLuc protein which produces a measurable luminescence signal. When the HiBiT-POI fusion is heated, as the protein aggregates and misfolds, the HiBiT tag cannot properly complement the LgBiT which results in reduced luminescence signal production. The smaller HiBiT tag has limited effect on thermostability and steric hindrance compared to the full-length NanoLuc protein.

Here, we describe the HiBiT CETSA protocol to measure target engagement in a cellular environment—a quantitative, rapid, and high-throughput way of screening many compounds. In overview, cells are transfected with HiBiT fusion protein plasmid, treated with compounds for screening, and heated across a temperature gradient to induce protein denaturation. Then cells are lysed with detergent, the remaining HiBiT signal is measured by providing complementary LgBiT and NanoLuc substrate, and the resulting luminescence signal is analyzed to determine if a compound has altered the proteins thermostability.

2 Materials

2.1 Cloning

1. Acceptor plasmids: pBiT3.1-N (Promega, #N2361) and pBiT3.1-C (Promega, #N2371).
2. Restriction enzymes and requisite digestion buffers, stored at – 20 °C.
3. Nuclease-free water.
4. Agarose gel (1%) with SYBR-Safe stain (or other DNA stain).
5. 1 kb DNA ladder.
6. Gel electrophoresis apparatus.
7. Gel purification kit (e.g., Qiagen, or Machery-Nagel).
8. Protein-of-interest (POI) primers are as follows:
   - *N-terminal* includes a Kozak sequence in the 5′ primer, includes a stop codon in the 3′ primer, and includes any additional necessary bases to maintain the reading frame.
   - *C-terminal* includes a ATG start codon in the 5′ primer and includes any additional necessary bases to maintain reading frame.
   - Depending on cloning method, remember to include any other required features to the primers.
9. Template DNA.
10. High-fidelity taq polymerase (e.g., NEB Q5, Thermo Fisher Phusion, etc.), stored at −20 °C.
11. PCR thermocycler.
12. Takara In-Fusion HD EcoDry Cloning reaction kit (stored in vacuum with desiccant).
13. NEB C3040 competent cells or other high transformation efficiency and reduced recombination chemically competent cells (e.g., Takara Stellar cells), stored at −80 °C and thawed on ice.
14. Water bath at 42 °C.
15. *SOC medium*: 2% tryptone, 0.5% yeast extract, 10 mM NaCl, 2.5 mM KCL, 10 mM $MgCl_2$, 10 mM $MgSO_4$, and 20 mM glucose.
16. *LB broth*: 10 g of tryptone, 5 g of yeast extract, and 10 g of NaCl per liter.
17. LB agar plates with appropriate antibiotic.
18. Kanamycin.
19. Sterile glass beads or sterile spreaders.
20. Incubator at 37 °C.
21. Round-bottom culture tubes with cap.

22. Shaker incubator at 37 °C.
23. Miniprep kit (e.g., Qiagen, or Machery-Nagel).
24. NanoDrop or other spectrophotometer for DNA quantification.
25. Sequencing primers for N- and C-HiBiT plasmids (e.g., T7 promoter).

2.2 Cell Culture

1. Any cell lines, e.g., HEK293T cells (ATCC #CRL3216) or HeLa cells (ATCC #CCL2) are used here.
2. *Growth medium*: DMEM (multicell) supplemented with 10% FBS, penicillin (100 U/mL), and streptomycin (100 μg/mL), stored at 4 °C and warmed to 37 °C prior to use.
3. Tissue culture-treated flasks or 10-cm plates.
4. Tissue culture incubator at 37 °C, 95% humidity, and 5% CO_2 concentration.
5. 1x Phosphate buffer saline.
6. 0.25% trypsin/EDTA, stored at 4 °C and warmed to 37 °C prior to use.
7. Centrifuge.
8. Cell counter or hemocytometer.
9. HiBiT-POI fusion DNA (generated by cloning or synthesized), stored at −20 °C.
10. pCDNA3.1 carrier DNA or any empty vector can be used, stored at −20 °C.
11. Opti-MEM reduced serum Medium or serum-free medium (Gibco), stored at 4 °C and warmed to 37 °C prior to use.
12. Any transfection reagent, e.g., X-tremeGENE HP transfection reagent (Sigma), or Fugene (Promega), stored at −20 °C and warmed to RT and vortexed gently prior to use.
13. Tissue culture-treated 6-well plates.

2.3 Compounds

1. DMSO or other solvent.
2. Compounds dissolved in DMSO or other solvent, stored at −20 °C.

2.4 HiBiT CETSA Assay

1. *Assay medium*: phenol red-free Opti-MEM Reduced Serum Medium (Gibco), stored at 4 °C and warmed to 37 °C prior to use (*see* **Note 1**).
2. Any 96-well PCR plates.
3. 96-well breathable film cover.
4. *Nano-Glo® HiBiT Lytic solution (Promega):* add 8 ml of Nano-Glo® HiBiT Lytic Buffer to a 15-ml centrifuge tube and add

80 μl of LgBiT Protein and 160 μl of Nano-Glo® HiBiT Lytic Substrate mixed by inversion.

5. 96-well plate adhesive cover compatible with your specific PCR plates and thermocycler (*see* **Note 2**).
6. Thermocycler with heat gradient, e.g., VeritiPro™ Thermal Cycler (Applied Biosystems), CFX96 Touch Real-Time PCR Detection System (Bio-Rad).
7. 384-well white microplates.
8. Microplate reader with luminescence detection, e.g., CLARIOstar (BMG), PHERAstar (BMG).
9. *Permeabilized CETSA assay buffer*: 1 tablet Protease Inhibitor Cocktail cOmplete (Roche), 1 tablet Phosphatase Inhibitor Cocktail (Roche) tablet dissolved in 10 mL of phenol red-free Opti-MEM Reduced Serum Medium.
10. *Cell permeabilization solution*: Dilute 50 mg/mL digitonin (Sigma #D141) in DMSO to make a 1000x stock. Store at RT and use at a final concentration of 50 μg/mL.

3 Methods

3.1 Generation of HiBiT Plasmids

For cloning one or a couple POIs, restriction enzyme-based cloning can be used to generate HiBiT fusions; however, for higher throughput and many targets, the use of Takara In-Fusion cloning is suggested. Both approaches can utilize the same primers. The two options for cloning mentioned here are just examples; any preferred cloning method can be utilized.

3.1.1 Preparation of Restriction Enzyme-Digested HiBiT Acceptor Plasmids

1. Take 2 μg of pBiT3.1-N or pBiT3.1-C and digest with 10 U of each enzyme (*see* pBiT3.1-N and pBiT3.1-C vector maps to see restriction sites available in the multiple cloning site, avoiding any that are present in the POI) and appropriate buffer in a 50 μl reaction volume for 1 h at 37 °C. Heat inactivate enzymes at appropriate temperature for 20 mins if using heat inactivation compatible enzymes.
2. Load DNA ladder, digested plasmid, and control uncut plasmid onto a 1% agarose gel (with SYBR DNA stain), and run electrophoresis until DNA ladder bands are well separated. Confirm that the plasmid was properly digested and cut out the linearized plasmid gel band and perform gel extraction according to manufacturer's protocol.

3.1.2 Preparation of POI Insert and Cloning into a HiBiT Acceptor Plasmid

1. Amplify and PCR clean-up (as per manufacturer's protocol) POI using primers that introduce the requisite restriction enzyme cut sites utilized to prepare the HiBiT acceptor

plasmids in **steps 1** and **2**, or In-Fusion specific primers (*see* **Note 3**). Alternatively, POI insert can be ordered as a DNA synthesis product.

2. Insert POI:

   (a) *By Restriction Enzyme*: First digest the POI PCR product using 10 U of each appropriate restriction enzymes in matching digestion buffer (use the same ones as selected in **step 3** for primer design). Ensure restriction enzymes are compatible with double digest or use serial digest if not compatible) in a 50 μl reaction volume incubated at 37 °C for 1 hr. PCR clean-up (as per manufacturer's protocol). Then, in a 10 μl reaction, add 5 μl 2x T4 ligation buffer, 1 μl T4 ligase enzyme, an appropriate amount of restriction enzyme-digested POI insert and acceptor plasmid (DNA amounts will depend on manufacturer's instructions for preferred T4 ligase reagent), and top up with nuclease-free water to reach 10 μl. Incubate for 10 mins–2 h at RT, or overnight at 16 °C. Heat inactivate at 65 °C for 10 mins and place on ice or freeze until use.

   (b) *By In-Fusion Reaction*: In a 10 μl reaction, add an appropriate amount of POI insert and restriction enzyme-digested acceptor plasmid as per manufacturer's manual (Takara), mix well, and add to an In-Fusion EcoDry lyophilized enzyme pellet. Resuspend well by pipetting repeatedly without introducing bubbles, and incubate at 37 °C for 15 min, followed by 50 °C for 15 min. Upon completion of reaction, place on ice.

3. Thaw chemically competent cells on ice. Add no more than 5 μl of ligation reaction of In-Fusion reaction to 50 μl of competent cells and incubate on ice for 30 min. Heat shock cells at 42 °C for 60 s using a thermocycler or water bath, recover on ice for 5 mins, add 450 μl pre-warmed SOC media, and then, incubate on a 37 °C for 1 hr. After recovery, spread 100 μl of bacterial culture on pre-warmed 10 cm LB agar plates containing kanamycin (50 μg/ml) using glass beads or sterile spreader. Incubate at 37 °C overnight.

4. Pick a few colonies, amplify each individual colony in LB containing appropriate antibiotic (e.g., kanamycin (50 μg/ml) for pBiT3.1-N and pBiT3.1-C (Promega)) in a round-bottom culture tube overnight in a shaker incubator at 37 °C. Purify plasmids from cultures using a miniprep kit as per manufacturer's protocol. Measure DNA concentration using a spectrophotometer, and sequence using appropriate primers to confirm positive inserts of POI.

3.2 HiBiT CETSA Assay Setup

The HiBiT CETSA assay is designed to evaluate target engagement of small molecules in a cellular environment through specific ligand-binding-induced protein stabilization over a temperature gradient as analyzed by NanoLuc (HiBiT+LgBiT). The HiBiT CETSA assay can be performed in live cells or in permeabilized cells. In live cells, the HiBiT CETSA assay can be used to evaluate the ability of a small molecule to enter a cell and engage a target protein in a native cellular environment. The permeabilized cell format CETSA enables screening of compounds with limited or unknown permeability and determines target protein engagement in a cellular environment.

3.2.1 Transient Transfection of Cells with HiBiT Fusion Protein

1. Grow HEK293T or HeLa in appropriate growth medium in tissue culture incubator at 37 °C, 95% humidity, and 5% CO_2 concentration.
2. Once cells have reached desired density, remove medium by aspiration, wash with PBS, and then add trypsin and incubate for at most 5 mins at 37 °C. Neutralize trypsin with media containing FBS. Spin cells down at 200× *g* for 5 minutes to pellet. Aspirate medium and resuspend cells in appropriate growth medium, pipetting up and down to break up cells. Use cell counter or hemocytometer to count cells and dilute to appropriate density for transfection.
3. *Prepare the Transfection DNA Mix*: Add 0.2 μg of N- or C-terminally tagged POI-HiBiT fusion DNA and 1.8 μg of empty plasmid (pCDNA3.1 used here) in 200 μl Opti-MEM and mix thoroughly. Add 4 μl X-tremeGENE HP transfection reagent, mix by inverting 5–10 times, and incubate at RT for 10–20 min to allow complexes to form (*see* **Note 4**).
4. Perform transfection as follows:
   (a) *By Forward Transfection*: Adjust HEK293T cell density to 2×10^5/ml (1×10^5/ml for HeLa cells) and plate cells in 6-well plates (2 ml/well). Grow for 24 h (70–90% confluency) and perform transfection by adding transfection DNA mix dropwise evenly over cells.
   (b) *By Reverse Transfection*: Adjust HEK293T cell density to 4×10^5/mL (2×10^5/ml for HeLa cells), plate cells in 6-well plates (2 ml/well), immediately add 200 μl incubated transfection DNA mix dropwise, and gently mix to distribute cells evenly (*see* **Note 5, 6,** and 7)0.5. The next day, proceed to Subheading 3.2.2 for Live-Cell CETSA, or Subheading 3.2.3 for Permeabilized Cell CETSA protocol.

3.2.2 Live-Cell Format Nano-Glo Cellular Thermal Shift Assay Protocol

1. Prepare diluted test compound(s) at 100X final concentration in 100% DMSO or preferred solvent (*see* **Notes 8** and **9**).
2. Trypsinize cells, neutralize trypsin using growth medium, pellet cells at 200× *g* for 5 mins, remove media, and resuspend in Opti-MEM at 2×10^5/ml, pipetting up and down repeatedly to break up cells.
3. Divide cells into compound-treated and untreated (DMSO) groups, adding 10 μl of 100X compound or DMSO control per 1 ml of resuspended cells (*see* **Note 10**).
4. Aliquot cells into 50 μl/well in a 96-well PCR plate, cover with breathable film, and incubate for 1 h at 37 °C, 95% humidity and 5% CO_2 concentration. Include one well per intended temperature, compound, and concentration to be tested in the CETSA including control vehicle DMSO/solvent and control temperature 37 °C (*see* **Notes 11, 12,** and **13**).
5. Immediately prior to heating the PCR plates and taking luminescence measurements, prepare Nano-Glo® HiBiT Lytic solution. Transfer 8 ml of Nano-Glo® HiBiT Lytic Buffer to a 15-ml centrifuge tube and add 80 μl of LgBiT protein and 160 μl of Nano-Glo® HiBiT Lytic Substrate. Mix by inversion.
6. Replace the breathable film on the PCR plate with compatible PCR film.
7. Heat samples as follows: 1 min at (22 °C) and 3 mins at desired temperature gradient, followed by a cooling step to chill samples to RT (22 °C). Once samples have reached RT (22 °C, as monitored in real time), remove the plate from the thermocycler, remove the PCR film, and incubate at 3 min at RT (22 °C).

   (*see* **Notes 14** and **15**).
8. Transfer 10 μl of cells into each well of 384-plate in triplicates or quadruplicates. Add 10 μL per well of Nano-Glo® HiBiT Lytic mix to the cooled PCR plate creating a final 1X substrate solution. Incubate 10 mins at RT.
9. Read signal immediately on a microplate reader with 10-- 30 seconds orbital shaking cycle at 300 rpm prior to luminescence measurement (*see* **Note 16**).

3.2.3 Permeabilized Cell Format Nano-Glo Cellular Thermal Shift Assay Protocol

1. Prepare diluted test compound(s) at 100X final concentration in 100% DMSO or preferred solvent (*see* **Notes 8** and **9**).
2. Trypsinize cells, neutralize trypsin using growth medium, pellet cells at 200× *g* for 5 mins, remove media, and resuspend in pre-warmed permeabilized CETSA assay buffer, pipetting up and down repeatedly to break up cells.
3. Prepare cell permeabilization solution by diluting the 50 mg/ ml digitonin stock solution 1:100 in phenol red-free Opti-

MEM Reduced Serum Medium to create a 10X stock solution (500 ug/ml).

4. Add 11.1 μL of 100X diluted compound to 1 ml resuspended cell suspension.
5. Add 5 μL per well of 10X digitonin solution (final in well concentration 50ug/ml) to a 96-well PCR plate.
6. Aliquot 45 μl/well cells into a 96-well PCR plate, cover with breathable film, and incubate for 1 h at 37 °C, 95% humidity and 5% CO_2 concentration. Include one well per intended temperature, compound, and concentration to be tested in the CETSA including control vehicle DMSO/solvent and control temperature 37 °C (*see* **Notes 11, 12, 13,** and **17**).
7. Follow **steps 5–10** as in Subheading 3.2.2.

3.2.4 Data Processing and Analysis of HiBiT CETSA Data

1. To generate apparent T_{agg} curves, the data are first converted to percent stabilized by relating the luminescence at temperature X to the luminescence at the lowest temperature for that given sample (e.g., using 37 °C as the reference). *See* **Note 19**.

$$\left(\frac{\text{luminescence } at \text{ X}^\circ\text{C for sample } A}{\text{luminescence } at \text{ } 37^\circ\text{C for sample } A}\right) \times 100 = \%\text{stabilized at } 37^\circ\text{C}$$

2. Data are then fitted to obtain apparent T_{agg} values using the nonlinear curve fit (e.g., Boltzmann Sigmoid equation) using GraphPad Prism software (*see* **Note 18**).
3. T_{agg} shifts (ΔT_{agg}) can be obtained by calculating the difference between T_{agg} DMSO-treated curve and compound-treated curves.

3.3 Considerations for Experimental Design

3.3.1 Comparison of N-Versus C-Terminally Tagged HiBiT Protein

Some proteins can show differences in thermodynamics in response to N- versus C-HiBiT tagging. It is advised to test both configurations and determine which is the most reproducible and stable. For instance, the melt curves for N-terminal HiBiT stabilize USP21 to higher temperatures than a C-tag, and thus, N- versus C-tagging may yield different results when testing compounds (Fig. 1).

3.3.2 Isothermal Compound Screening

In some cases, screening many compounds at a single temperature makes more sense and can also allow for higher throughput. In isothermal screens, it is paramount to empirically identify an optimal temperature based on the largest T_{agg} shift (ΔT_{agg}) from comparing melting curves (Figs. 2a, 3a, b, and 4a). Such isothermal screens are even useful for difficult targets like membrane-associated proteins like three-prime repair exonuclease 1 protein (TREX1) that have high melting temperatures (Fig. 3c). Once an optimal temperature is calculated, additional compounds can be screened using this one temperature and compounds at varying

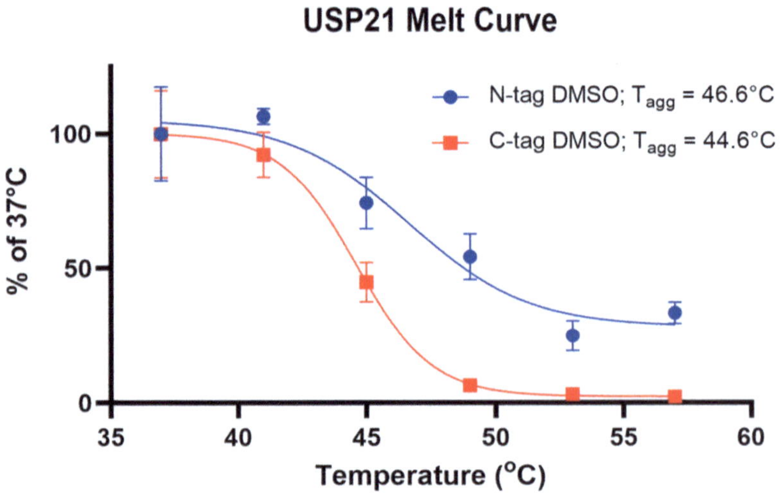

Fig. 1 Comparison of N- versus C-terminally tagged USP21. HEK293T cells were transfected with either HiBiT-USP21 or USP21-HiBiT. After 24 h, cells were treated with DMSO (0.1% final concentration) for 1 h, heated in a thermal cycler at temperatures 37–57 °C, lysed and luciferase substrate was added, and luminescence was measured. Data points represent the intensity of luminescence as a % of the 37 °C treatment ± SD ($n = 4$ technical replicates)

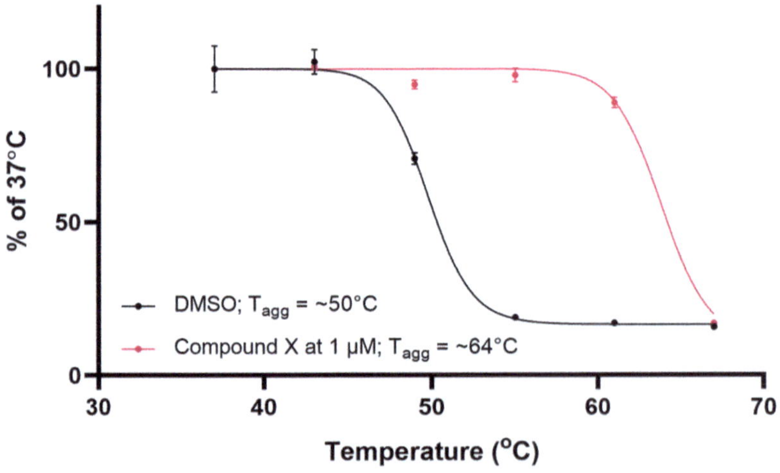

Fig. 2 Dose–response curves for protein Y with compound X CETSA. HEK293T cells were transfected with protein Y-HiBit. After 24 h, cells were treated with DMSO (0.1% final concentration) or compound X at various concentrations for 1 h, heated in a thermal cycler at temperatures 37–67 °C, lysed and luciferase substrate was added, and luminescence was measured. Data points represent the intensity of luminescence as a % of the 37 °C treatment ± SD ($n = 4$ technical replicates). Plotted is a comparison of the melting curve comparing thermostabilization between DMSO and compound-treated protein Y, yielding an $\sim\Delta T_{agg}$ of 14 °C between DMSO and compound X

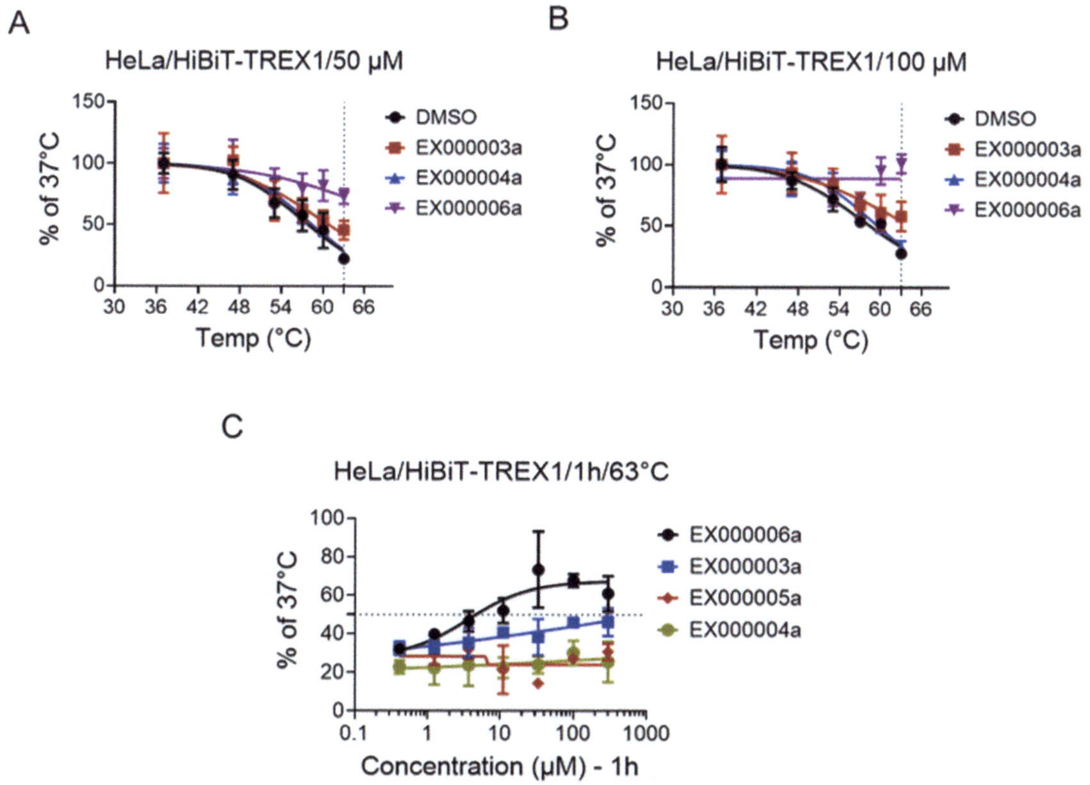

Fig. 3 Optimization of HiBiT CETSA assay conditions for TREX1. (**a**, **b**) HeLa cells were transfected with HiBiT-TREX1. After 24 h, cells were treated with DMSO or potential TREX1 chemical probes each at concentrations of 50 μM (A) or 100 μM (**b**) for 1 h, followed by heating in a thermal cycler at temperatures 37–63 °C. Cells were lysed, LgBiT and luciferase substrate were added, and luminescence was immediately measured. The highest thermal shift was observed with compound EX000006a at 100 μM. Data points represent the intensity of luminescence as a % of corresponding treatments at 37 °C ± SD ($n = 4$ technical replicates). (**c**) HeLa cells were transfected with HiBiT-TREX1. After 24 h, cells were treated with DMSO or compounds TREX1 chemical probes at concentrations from 0.4–300 μM for 1 h, followed by heating in a thermal cycler at a temperature of 63 °C. Cells were lysed, LgBiT and luciferase substrate were added, and luminescence was immediately measured. The highest thermal shift and potency were observed with compound EX000006a. Data points represent the intensity of luminescence as a % of corresponding treatments at 37 °C ± SEM ($n = 2$ independent experiments)

concentrations to successful identify stabilizing compounds (Fig. 3c).

A caution about isothermal compound screening is that while some small molecules do not show any significant protein stabilization at a given temperature and concentration, this does not mean it will not at other temperatures/concentration, and vice-versa. It is advisable that for hit compounds, more thorough screening at multiple temperatures and concentrations is also carried out to get the range of EC_{50}s across conditions (Fig. 4).

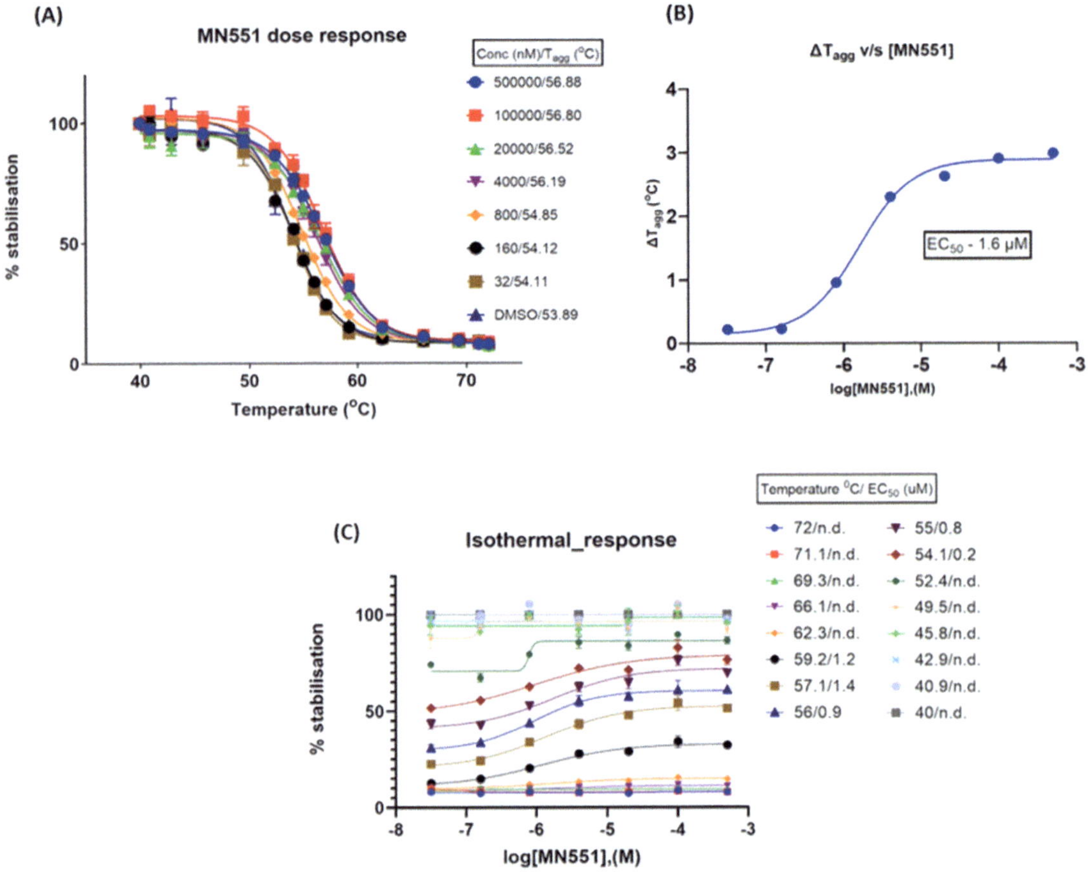

Fig. 4 Dose–response curves for MN551 CETSA. (**a**) MN551 mediated thermal stabilization of transfected HiBiT-SOCS2 fusion as observed in a dose-dependent format. (**b**) EC50 is obtained from the plot of shift in aggregation temperature (T_{agg}) with increasing concentrations of MN551. (**c**) Comparison of EC50 obtained from isothermal plots at different temperatures

Overall isothermal screens provide the ability to characterize many compounds while minimizing reagents utilized. Isothermal screens are particularly useful in scenarios where a reference compound or well-established positive control compound is compared to novel uncharacterized compounds or where a hit compound is compared against chemically optimized compounds within the same series to determine if there are any improvements in potency.

3.3.3 Dose–Response Compound Screening: Total Temperature Gradient v/s Isothermal Screening

Isothermal compound screening is the ideal method to screen multiple compounds in a high-throughput manner for their potency. However, the choice of temperature used for such screening is critical to successful ranking of the binders.

The traditional choice of a single screening temperature for diverse set of binders relies on the expectation that the binders would elicit similar T_{agg} changes in the POI. A ligand dose response coupled with a total temperature gradient consumes larger quantities of reagents and compounds but provides a higher confidence in the calculated EC50.

Here, we present an example involving monitoring the covalent engagement of the E3 ligase SOCS2 with the ligand MN551 (Fig. 4) [9]. The figure shows a comparison of EC50 derived from the dose responses with a total temperature gradient (Fig. 4a, b) and EC50 determined through isothermal dose response (Fig. 4c). Figure 4c highlights the importance of the choice of temperature in the reliable estimation of EC50. For optimal EC50 calculations, try to include concentrations that yield a plateau at either end of the dose–response curve (e.g., high and low ends).

3.4 Other Applications of HiBiT CETSA

HiBiT tag-based CETSA is also useful for screening prodrugs [9], such as the example of the HiBiT CETSA that was used to evaluate and rank the permeability efficacy of different prodrug versions of a binder (preprint publication). In this screening, we utilized the live-cell HiBiT CETSA to identify pivaloyloxymethyl (POM) protecting group as the most efficient masking group for MN551.

As the HiBiT-based technology continues to be developed, such as genome-editing to insert HiBiT tags [10], we anticipated novel applications for HiBiT CETSA will continue to emerge and accelerate drug discovery. For example, a logical extension is to utilize HiBiT CETSA in permeabilized versus live-cell format to compare EC50s and more broadly establish permeability of ligands akin to the NanoBRET target engagement assays [11]. Thus, HiBiT CETSA is a useful method to determine the compound target engagement in cells and drive the compound structure–activity relationship.

4 Notes

1. Assay medium can affect the results and can be target specific. FBS-free and antibiotic-free phenol red-free media, Opti-MEM Reduced Serum Medium, or serum-free phenol red-free medium is suggested if there is concern that compounds may interact with serum or antibiotics. This assay medium is recommended as a starting place, but other buffers may need to be tested to optimize assay conditions specific to your target.
2. Compatible PCR plates and plate adhesives may vary depending on your thermal cycler instrument; please check requirements for your specific machine.
3. For In-Fusion primers, the Takara In-Fusion Primer design tool is convenient to design primers to produce PCR

amplification inserts with selected restriction enzyme cut sites and/or to include the required 15 bp overlap with the acceptor plasmid on the 5′- and 3′-ends for a successful In-Fusion reaction.

4. Larger or smaller bulk transfections should be scaled accordingly, using this ratio.
5. This transfection protocol is for a 1:10 HiBiT fusion dilution into Carrier DNA dilution (e.g., 0.1 μg of POI-HiBiT fusion DNA is utilized for every 0.9 μg of carrier pCDNA3.1 DNA). It is recommended that you begin with testing 1:10 and 1:100 DNA dilution (e.g., 0.01 μg of POI-HiBiT fusion DNA is utilized for every 0.99 μg of carrier pCDNA3.1 DNA) on your target to establish the optimum expression level to observe T_{agg} shifts. Other expression levels may need to be tested to yield optimal apparent T_{agg} shifts for your target of interest.
6. Live-cell CETSA usually requires a higher concentration of HiBiT-POI fusion than the permeabilized format of the CETSA. For example, use a 1:10 dilution ratio for the live-cell CETSA, if you are using 1:100 ratio for the permeabilized cell CETSA.
7. It is recommended to test both N- and C-terminal fusions to determine the optimal orientation for each target protein.
8. Cellular thermal shift assays tend to require high amounts of compound to result in measurable apparent T_{agg} shifts; however, this may vary depending on the characteristics of the tested compounds. Thus, it is recommended to test several concentrations of test compounds to determine the amount needed to obtain a measurable apparent T_{agg} shift. If a positive control-binding compound exists for the target POI, consider using a concentration gradient to determine a range of compound concentrations that yield measurable T_{agg} shifts in other test compounds.
9. We suggest starting with testing final concentrations of 30–50 μM and above. The final concentration range can be adjusted after initial screening results.
10. Use matched final solvent concentration in control and treated samples (e.g., final DMSO 1%).
11. Optimization of incubation times may be required depending on the characteristics of the test compound and/or your target protein. 1–4 hr incubations are typically used as a starting point.
12. We suggest a minimum of six temperature points starting from 37 °C and increasing in 5 °C increments to 67 °C for initial testing before getting an idea of the stability of the tagged

protein. However, what is feasible will depend on specific thermocyclers, so adjust the number of temperature points and increments for your equipment, keeping in mind that more points and a larger temperature span are beneficial for initial screening. Final temperature ranges will vary by protein, but this initial broad temperature range testing will allow experimenters to get an idea of the stability of their protein and adjust temperature ranges as needed.

13. To increase the number of temperature points tested in a gradient, experiments can be split between multiple PCR plates, e.g., one covering 37–55 °C, a second covering 55–73 °C. Splitting the temperature gradient between two thermocycler runs using two separate PCR plate setups allows sensitive detection of minor changes in T_{agg}.
14. Setting the thermocycler to 22 °C before placing the PCR plate inside protects against noise introduced by thermocyclers with resting temperatures that are higher than 37 °C or may still be warm from previous use. Total heating times and temperature gradient spans may vary depending on your target as well as your thermal cycler. It is recommended to start with a 3 mins heating time and adjust to longer or shorter heating times depending on the shape of the melting curve.
15. The inclusion of the cooling to RT step in the thermocycler ensures that sample temperatures all get cooled to RT within the shortest timeframe, preventing further denaturing and any nonuniform cooling between samples exposed to varying temperatures. An ice block can also be placed on top of the plate once removed from the thermocycler to prevent any heat transfer from the heated plate top to samples at the bottom of the PCR wells.
16. Adjust the gain to 90–95% of the well with the highest signal and take a second reading. This is important to set an appropriate assay window and sensitivity to avoid signal saturation. For instance, if the signal from a well is low, but the gain is set too low, then it will be difficult to distinguish signal from background noise, whereas if a gain is set too high, wells with high signal will saturate the instrument and the data become unusable.
17. This addition of compound–cell mixture to digitonin in plates is important for proper mixing of cells with the digitonin and test compounds. In-plate digitonin permeabilization as opposed to bulk sample permeabilization yields more consistent results as some targets and cell lines are sensitive to bulk permeabilization.
18. It is recommended that samples should be run in biological replicates at a minimum of $n = 2$ to 4 and technical replicates of

$n = 3$ to 4. Ideally, experiments should be performed at least three times, each on different days, to get statistically meaningful results.

19. In some instances, a significant difference between compound-treated and vehicle controls at 37 °C can be observed. Some compounds may increase the luciferase signal by stabilizing proteins at 37 °C; however, if the compound treatment results in the suppression of luciferase signal, it may indicate compound toxicity or an indirect effect on luciferase enzyme activity. The interpretation of results should consider possible melting curve artifacts due to signal changes. Therefore, we recommend signals to be normalized to unheated, compound-treated controls as well as the inclusion of an unrelated HiBit-tagged protein control which is not stabilized by compound treatment. Additionally, to reduce compound-induced luciferase signal suppression caused by nonspecific effects on luciferase activity, samples after lysing (Subheading 3.2.2, **step 7**) can also be diluted before substrate addition by twofold to ten-fold depending on signal levels.

Acknowledgements

This research was supported by the Natural Sciences and Engineering Research Council of Canada through a postdoctoral fellowship to V.V. and grant to D.B.L, and by the Structural Genomics Consortium, a registered charity (no: 1097737) that receives funds from Bayer AG, Boehringer Ingelheim, Bristol Myers Squibb, Genentech, Genome Canada through Ontario Genomics Institute [OGI-196], EU/EFPIA/ OICR/McGill/KTH/Diamond Innovative Medicines Initiative 2 Joint Undertaking [EUbOPEN grant 875510], Janssen, Merck KGaA (aka EMD in Canada and US), Pfizer, and Takeda. This project has also received funding from the Innovative Medicines Initiative 2 (IMI2) Joint Undertaking (JU) under grant agreement No 875510. The JU receives support from the European Union's Horizon 2020 research and innovation program and EFPIA and Ontario Institute for Cancer Research, Royal Institution for the Advancement of Learning McGill University, Kungliga Tekniska Hoegskolan, Diamond Light Source Limited. S.R. is specifically funded by the IMI2 EUbOPEN project. S. H.B was supported by Mitacs Elevate Postdoctoral Fellowship.

References

1. Simon GM, Niphakis MJ, Cravatt BF (2013) Determining target engagement in living systems. Nat Chem Biol 9:200–205. https://doi.org/10.1038/NCHEMBIO.1211
2. Molina DM, Jafari R, Ignatushchenko M et al (2013) Monitoring drug target engagement in cells and tissues using the cellular thermal shift

assay. Science 341:84–87. https://doi.org/10.1126/SCIENCE.1233606

3. Jafari R, Almqvist H, Axelsson H et al (2014) The cellular thermal shift assay for evaluating drug target interactions in cells. Nat Protoc 9: 2100–2122. https://doi.org/10.1038/NPROT.2014.138
4. Hall MP, Unch J, Binkowski BF et al (2012) Engineered luciferase reporter from a deep sea shrimp utilizing a novel imidazopyrazinone substrate. ACS Chem Biol 7:1848–1857. https://doi.org/10.1021/CB3002478
5. Dart ML, Machleidt T, Jost E et al (2018) Homogeneous assay for target engagement utilizing bioluminescent thermal shift. ACS Med Chem Lett 9:546–551. https://doi.org/10.1021/ACSMEDCHEMLETT.8B00081/SUPPL_FILE/ML8B00081_SI_001.PDF
6. Sanchez TW, Owens A, Martinez NJ et al (2021) High-throughput detection of ligand-protein binding using a SplitLuc cellular thermal shift assay. Methods Mol Biol 2365:21–41. https://doi.org/10.1007/978-1-0716-1665-9_2
7. Martinez NJ, Asawa RR, Cyr MG et al (2018) A widely-applicable high-throughput cellular thermal shift assay (CETSA) using split Nano luciferase. Sci Reports 81(8):1–16. https://doi.org/10.1038/s41598-018-27834-y
8. Mortison JD, Cornella-Taracido I, Venkatchalam G et al (2021) Rapid evaluation of small molecule cellular target engagement with a luminescent thermal shift assay. ACS Med Chem Lett 12:1288–1294. https://doi.org/10.1021/ACSMEDCHEMLETT.1C00276/SUPPL_FILE/ML1C00276_SI_003.XLSX
9. Ramachandran S, Makukhin N, Haubrich K, et al (2022) Structure-based design of a phosphotyrosine-masked covalent ligand targeting the E3 ligase SOCS2. https://doi.org/10.26434/CHEMRXIV-2022-BVJ80
10. Larson HG, Zakharov AV, Sarkar S et al (2021) A genome-edited ERα-HiBiT fusion reporter cell line for the identification of ERα modulators via high-throughput screening and CETSA. Assay Drug Dev Technol 19:539–549. https://doi.org/10.1089/ADT.2021.059
11. Vasta JD, Corona CR, Robers MB (2021) A high-throughput method to prioritize PROTAC intracellular target engagement and cell permeability using NanoBRET. Methods Mol Biol 2365:265–282. https://doi.org/10.1007/978-1-0716-1665-9_14

Chapter 12

Detection of Cellular Target Engagement for Small-Molecule Modulators of Striatal-Enriched Protein Tyrosine Phosphatase (STEP)

Ye Na Han, Lester J. Lambert, Laurent J. S. De Backer, Jiaqian Wu, Nicholas D. P. Cosford, and Lutz Tautz

Abstract

Striatal-enriched protein tyrosine phosphatase (STEP) is a brain-specific enzyme that regulates the signaling molecules that control synaptic plasticity and neuronal function. Dysregulation of STEP is linked to the pathophysiology of Alzheimer's disease and other neuropsychiatric disorders. Experimental results from neurological deficit disease models suggest that the modulation of STEP could be beneficial in a number of these disorders. This prompted our work to identify small-molecule modulators of STEP to provide the foundation of a drug discovery program. As a component of our testing funnel to identify small-molecule STEP inhibitors, we have developed a cellular target engagement assay that can identify compounds that interact with $STEP_{46}$. We provide a comprehensive protocol to enable the use of this miniaturized assay, and we demonstrate its utility to benchmark the binding of newly discovered compounds.

Key words STEP, PTPN5, Cellular target engagement assay, CETSA, Cellular thermal shift, InCELL Pulse, Protein tyrosine phosphatase, Small molecule, Inhibitor, Neurodegenerative disorders, Alzheimer's disease, Drug discovery, Protein–drug interaction

1 Introduction

Striatal-enriched protein tyrosine phosphatase (STEP), encoded by the *PTPN5* gene, is a neuron-specific protein tyrosine phosphatase (PTP) that opposes the development of synaptic strengthening [1, 2]. High levels of active STEP contribute to the cognitive deficits in various neurodegenerative and neuropsychiatric disorders, including Alzheimer's disease (AD) [3], Parkinson's disease [4], schizophrenia [5], and fragile X syndrome [6]. Interestingly, STEP knockout (KO) mice show enhanced memory and learning abilities [7, 8]. Moreover, genetic reduction in STEP in mouse

Daniel Merk and Apirat Chaikuad (eds.), *Chemogenomics: Methods and Protocols*, Methods in Molecular Biology, vol. 2706, https://doi.org/10.1007/978-1-0716-3397-7_12,

models of AD, schizophrenia, and fragile X syndrome reverses the cognitive and cellular deficits typically present in these models [5, 6, 9]. Those data suggest that small-molecule inhibitors of STEP could be beneficial for the treatment of AD and other neurological disorders. Indeed, a previously identified STEP inhibitor, TC-2153, was able to phenocopy the effects of STEP KO in a mouse model of AD [10]. While TC-2153 has served as a useful tool compound in multiple studies [11–14], its covalent and oxidative mechanism of action [10] and its potential to react with cellular thiols and modify DNA [15–18] have precluded this inhibitor from further preclinical advancement. Other reported STEP inhibitors suffer from poor selectivity for STEP and/or lack of efficacy under physiological conditions [19–22].

As part of our discovery platform to develop novel STEP inhibitors with improved properties, we have developed a cellular thermal shift assay (CETSA) protocol to assess the target engagement of candidate compounds in cells [23]. CETSA has found use as reliable means to validate and quantify small-molecule lead compound interactions in cells [24]. CETSA is based on the principle that the binding of a small molecule to a target protein can change its thermal stability. The original CETSA protocol uses heat pulses at varying temperatures and immunoblotting to quantify intact target protein [25]. However, several reporter-based systems that rely on the heterologous expression of targets have been developed and provide the potential for miniaturization and higher throughput [26, 27]. One such system is the InCELL Pulse™ platform (Eurofins DiscoverX). We previously utilized InCELL Pulse to study target engagement of small-molecule inhibitors with oncogenic forms of the SHP2 phosphatase [28, 29]. Based on this prior success, we have used InCELL Pulse to develop a cellular target engagement assay for $STEP_{46}$, one of the two major splice variants of STEP.

InCELL Pulse is based on a β-galactosidase enzyme fragment complementation (EFC) assay (Fig. 1) [30]. The protein of interest is expressed as an N- or C-terminal fusion protein with an enhanced ProLabel® tag (ePL), a 42 amino acid fragment of β-galactosidase. After cell treatment with a candidate compound, application of a heat gradient, and cell lysis, a reporter enzyme acceptor (EA) is added, resulting in detectable β-galactosidase activity. Due to the applied temperature gradient, proteins will denature and aggregate as the temperature increases based on their thermal stability. The binding of a small molecule can stabilize (or destabilize) the target protein, and this change in thermal stability is quantified, as the ePL tag is only available for complementation when the target protein is intact and in solution. Melting curves are recorded for both candidate compound and vehicle treatment. A significant change in

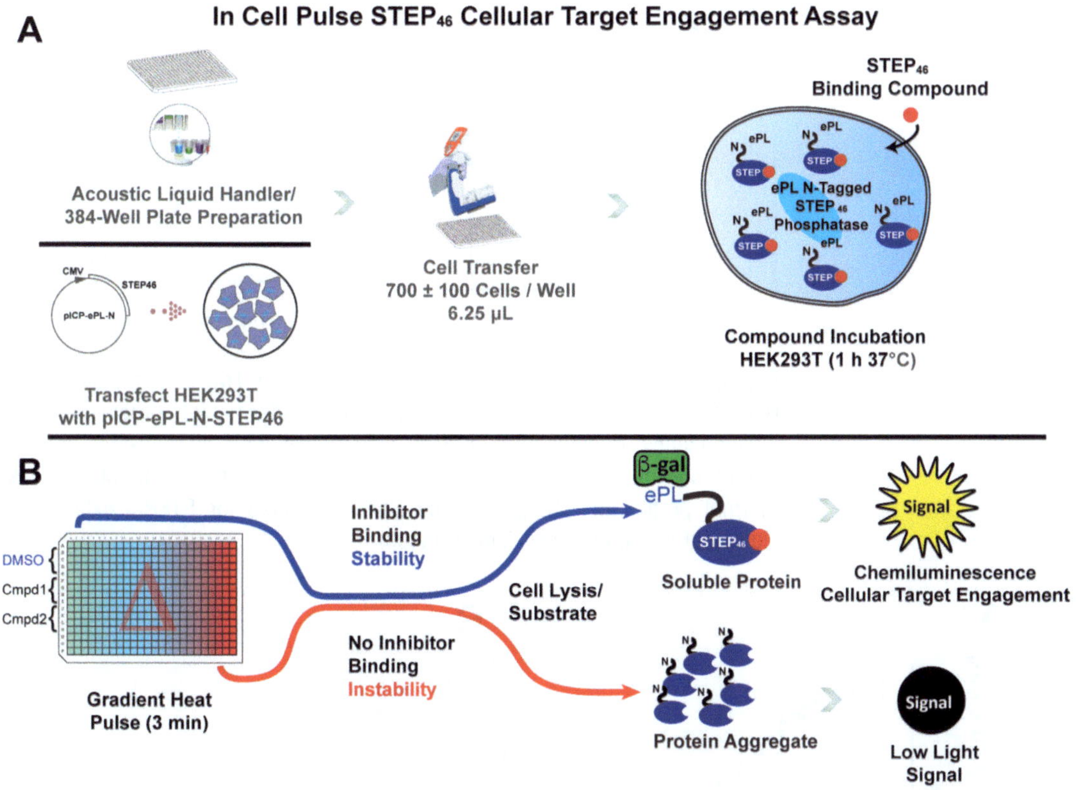

Fig. 1 Principles and Workflow of the InCELL Pulse $STEP_{46}$ Cellular Target Engagement Assay. The miniaturized $STEP_{46}$ target engagement assay is a form of reporter-based cellular thermal shift assay that can be reliably integrated into a drug discovery campaign. (**a**) An assay plate is prepared using an Echo Liquid Handler (or similar) to spot compounds of interest in the wells to be probed. HEK293T cells are transiently transfected with a pICP-ePL-N-STEP46 plasmid that expresses $STEP_{46}$ with an enhanced ProLabel (ePL, 42 amino acids) fusion tag. After 1 d, cells are detached, resuspended in fresh growth media, and are transferred to the assay plate. The cells are incubated with candidate compounds for 1 h. (**b**) For a thermal profile, the assay plate is subjected to a temperature gradient pulse for 3 min. Increasing temperatures cause proteins to denature and form insoluble, inaccessible aggregates. Specific target engagement of a small molecule can stabilize $STEP_{46}$ against aggregation. A mixture of enzyme acceptor (EA) complementation reagent and lysis buffer enables the quantification of soluble ePL-tagged $STEP_{46}$ via the reporter enzyme chemiluminescence system. The signal for each well is recorded and the data are analyzed

melting temperature (T_m) is indicative of cellular compound binding to the target protein. The CETSA protocol for $STEP_{46}$ provided below utilizes advanced instrumentation for compound acoustic dispensing. However, the assay can be adapted to existing equipment in most laboratories. An example of assay performance for $STEP_{46}$ is shown in Fig. 2. We have successfully used this protocol to confirm target engagement of novel STEP inhibitors developed in our laboratory (Fig. 3).

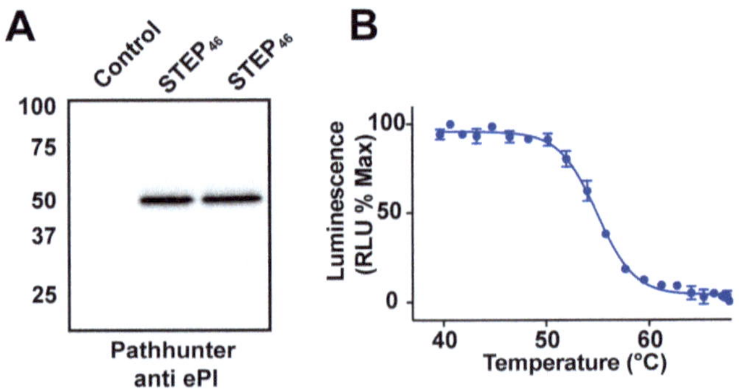

Fig. 2 $STEP_{46}$ Expression and Thermal Profile. (**a**) Transient transfection of HEK293T cells with pICP-ePL-N-STEP46. Western blot probing of two independently transfected HEK293T plate wells with an anti-EPL antibody (PathHunter) shows reliable expression of the $STEP_{46}$ protein. (**b**) InCELL Pulse thermal profile for $STEP_{46}$. The cellular melt curve for $STEP_{46}$ exhibits the sigmoidal shape of a well-folded protein

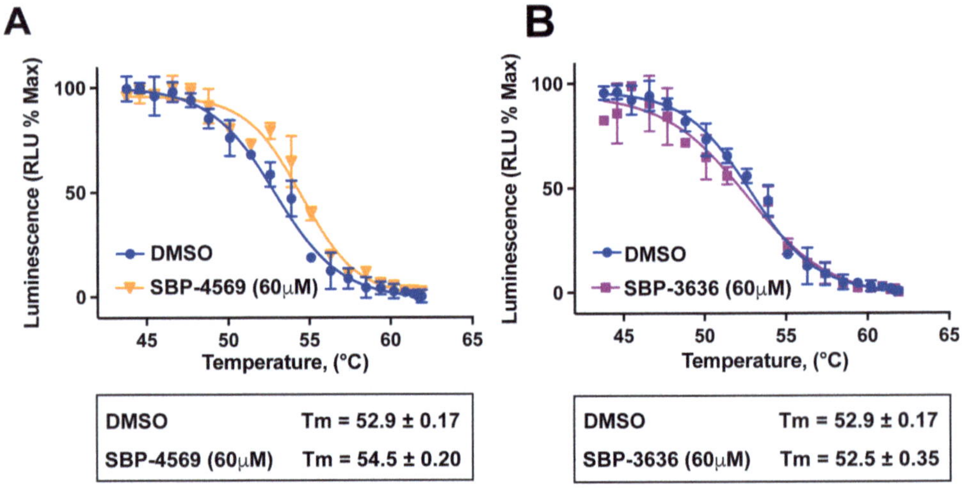

Fig. 3 Small-Molecule Probesf of $STEP_{46}$ can Stabilize the Enzyme (**a**) Cellular target engagement profiles of $STEP_{46}$ with small-molecule inhibitor SBP-4569 (orange) and vehicle control (DMSO, blue). The stabilization of $STEP_{46}$ by SBP-4569 is evident with adequate sampling of the transition temperature. (**b**) InCELL Pulse thermal profiles of $STEP_{46}$ with small-molecule inhibitor SBP-3636 (magenta) and vehicle control (DMSO, blue). Although chemically similar to SBP-4569, SBP-3636 does not produce a shift of the $STEP_{46}$ melting temperature, suggesting that this compound either does not enter cells or does not bind to STEP in cells

2 Materials

Measurement of cellular target engagement for compounds using the InCELL Pulse target engagement assay utilizes common commercial sources for the components and reagents. Storage conditions and specific handling measures are noted.

2.1 Cell Culture

1. HEK293T cells (ATCC).
2. jetPRIME® reagent and buffer (Polyplus, Illkirch, France).
3. TrypLE™ Express (Gibco/Thermo Fisher).
4. 6-well tissue culture (TC)-treated cell culture plates.
5. Growth media: Dulbecco's modified Eagle's medium (DMEM 1X + GlutaMAX™; Gibco/Thermo Fisher; 500 mL), fetal bovine serum (58 mL; 10%), 100X antimycotic–antibiotic (5.8 mL; 1X), 1 M HEPES (11.2 mL; 20 mM), and 100 mM sodium pyruvate (5.8 mL; 1 mM). Store at 4 °C.

2.2 Assay Components

1. InCELL Pulse Starter Kit (DiscoverX, Eurofins). Make 1 mL aliquots of the three assay components (EA reagent, lysis buffer, and substrate) and store at −20 °C.
2. ePL-tagged expression plasmid for $STEP_{46}$: Prepare by PCR amplification of the $STEP_{46}$ gene, digestion with restriction enzymes EcoRI and XbaI, and directional cloning into plasmid pICP-ePL-N. Propagate the pICP-ePL-N-STEP46 plasmid using *E. coli* strain DH5α and the GeneJET Plasmid Maxi Prep Kit (Thermo Fisher).

2.3 Instrumentation

1. Mastercycler X50h 384-Well Gradient-Capable Thermal Cycler (Eppendorf).
2. Echo® 555 Liquid Handler (Labcyte).
3. Tecan SPARK® Multimode Microplate Reader (Tecan).
4. E1-ClipTip™ Multichannel Pipette (Thermo Fisher).
5. 384-Well Low Dead Volume (LDV) Echo-Qualified Plates (Labcyte).
6. Armadillo High-Performance 384-Well White PCR Plates (Thermo Fisher).
7. Countess™ II FL Automated Cell Counter (Thermo Fisher).

3 Methods

3.1 Cell Culture and Transient Transfection with Target Engagement Plasmid

1. Revive HEK293T cells preserved in cryo-storage and maintain adherent cells in growth media in a low passage state at 37 °C, 5% CO_2.
2. Split cells bi-weekly. Do not utilize cells that have been passaged greater than 25 times to ensure assay reproducibility.
3. Detach HEK293T cells from a 75-cm^2 flask using TrypLE cell detachment solution (3 mL), dilute with growth media (12 mL), and collect the cells by centrifugation at 1400× *g* for 4 min. Resuspend cells in growth media (10 mL). Measure both the cell density and cell viability using Trypan blue and a

countess cell counter. Plate 0.7×10^6 HEK293T cells per well in a 6-well TC-treated cell culture plate (2 mL per well). Incubate for 24 h at 37 °C, 5% CO_2.

4. Dilute 2 μg DNA from a purified plasmid stock of pICP-ePL-N-STEP46 (500 ng/μL) into 200 μL of jetPRIME buffer. Vortex for 10 s, add 4 μL jetPRIME reagent, vortex, and incubate for 10 min at room temperature. Add the plasmid transfection mixture to one well of the 6-well cell culture plate containing 2 mL cells (*see* **Note 1**). Incubate at 37 °C, 5% CO_2 for 24 h.
5. Confirm expression levels of $STEP_{46}$ using immunoblotting and PathHunter® anti-PK/PL antibodies (DiscoverX, Eurofins) (Fig. 2a).

3.2 Cell Detachment and Assay Plate Preparation

1. Aspirate growth media and add 0.3 mL TrypLE cell detachment reagent to the adherent surface-bound cells. Incubate cells at 23 °C for 2 min. Add 1 mL of growth media to cells. Gently dislodge cells by pipet (3x) and transfer cells to a 15-mL Falcon tube. Centrifuge at 1400× *g* for 4 min. Aspirate media and replace with ~3 mL growth media. Measure the concentration and viability of cells with a cell counter. Dilute cells to 0.125×10^6 cell/mL and use cells within 2 h (*see* **Note 2**).
2. Spot compounds in quadruplicate into a 384-well Twin.tec 384 real-time PCR plate using an Echo Liquid Handler or equivalent (e.g., 20 nL of a 20 mM compound stock solution; *see* **Note 3**). Add cells to a sterile single-channel trough and add 6.25 μL cells to each assay well using a multichannel pipette. Centrifuge plate at 42× *g* for 30 s, apply a lid seal, and incubate the assay plate at 37 °C, 5% CO_2 for 1 h.
3. Prepare the InCELL Pulse Master Mix according to the manufacturer's protocol (EA-10; 3 mL Master Mix for one 384-well plate; volume fractions: substrate buffer (0.67), EA reagent (0.17), and lysis buffer (0.17)).

3.3 Thermal Pulse and Assay Quantification

1. Remove the assay plate from the incubator and apply a 3 min heat pulse using a gradient-capable thermal cycler with a desired temperature gradient (e.g., horizontal gradient of 42–62 °C across 24 wells). Employ a 15 s countdown to enable stable temperatures to be established when the plate is placed on the grid. (*see* **Note 4**). Add a recovery step of 3 min at 20 °C after the heat pulse has been applied.
2. Add to each assay well 6.25 μL of the InCELL Pulse Master Mix. Centrifuge plate at 42× *g* for 30 s and incubate at ambient temperature for 30–60 min.

3. Measure chemiluminescence with the use of a Tecan Spark Multimode microplate reader or equivalent instrument capable of reading chemiluminescence (integration time, 1000 ms; settle time, 0 s).
4. Analyze chemiluminescence data using GraphPad Prism or an equivalent program. Calculated curve fits are as follows: normalize chemiluminescent values with the maximum and minimum value defined as 100% or 0%, respectively. Apply a Boltzmann sigmoidal nonlinear least squared fit of the normalized chemiluminescence and calculate an EC_{50} value, which corresponds to the T_m value (*see* **Note 5**) (Fig. 2b).

4 Notes

1. Transfect from a single source of concentrated plasmid for uniform assay results. One well of transfected cells (0.7×10^6 cells) will be enough for two 384-well CETSA plates.
2. Cell viability (>90% survival) is a critical parameter that assures that meaningful data about cellular penetrance and target engagement are obtained. The one-hour incubation usually does not result in significant cell death for most compounds. Increasing the scale of the assay could be useful in adapting it to a high-throughput campaign. However, high cell viability should not be compromised.
3. If compound stock solutions are in DMSO, the amount of stock solution added should be chosen so that the final DMSO concentration is ≤0.5%.
4. As a depletion assay, meaningful data from a thermal profile would optimally be obtained by sampling more data points at the melt temperature. For $STEP_{46}$, we set the temperature gradient from 38 to 68 °C or 42 to 62 °C, with the narrow temperature span best able to assess the effect of target engagement.
5. A sigmoidal melt curve with a narrow transition temperature is observed for $STEP_{46}$. This is indicative of a well-folded protein in a cellular environment. Melt curves that exhibit broad, indiscrete transitions could indicate that the cellular protein is not correctly folded under the chosen expression conditions. Interpretation of melt curve shape due to compound binding has been described [31]. Optimization of the transfection protocol can be performed to modulate the expression level and can influence the observed temperature profile.

Acknowledgements

Research reported in this publication was supported by the National Institutes of Health under awards numbers R01AG065387 and R21AG067155 (to L. T.) and by the NCI Cancer Center Support Grant P30CA030199. The content is solely the responsibility of the authors and does not necessarily represent the official views of the National Institutes of Health.

References

1. Lombroso P, Murdoch G, Lerner M (1991) Molecular characterization of a protein-tyrosine-phosphatase enriched in striatum. Proc Natl Acad Sci U S A 88:7242–7246
2. Lombroso PJ, Ogren M, Kurup P et al (2016) Molecular underpinnings of neurodegenerative disorders: striatal-enriched protein tyrosine phosphatase signaling and synaptic plasticity. F1000Res 5
3. Xu J, Kurup P, Nairn AC et al (2012) Striatal-enriched protein tyrosine phosphatase in Alzheimer's disease. Adv Pharmacol 64:303–325
4. Kurup PK, Xu J, Videira RA et al (2015) STEP61 is a substrate of the E3 ligase parkin and is upregulated in Parkinson's disease. Proc Natl Acad Sci U S A 112:1202–1207
5. Carty N, Xu J, Kurup P et al (2012) The tyrosine phosphatase STEP: implications in schizophrenia and the molecular mechanism underlying antipsychotic medications. Transl Psychiatry 2:e137
6. Goebel-Goody S, Wilson-Wallis E, Royston S et al (2012) Genetic manipulation of STEP reverses behavioral abnormalities in a fragile X syndrome mouse model. Genes Brain Behav 11:586–600
7. Venkitaramani D, Paul S, Zhang Y et al (2009) Knockout of striatal enriched protein tyrosine phosphatase in mice results in increased ERK1/2 phosphorylation. Synapse 63:69–81
8. Venkitaramani D, Moura P, Picciotto M et al (2011) Striatal-enriched protein tyrosine phosphatase (STEP) knockout mice have enhanced hippocampal memory. Eur J Neurosci 33: 2288–2298
9. Zhang Y, Kurup P, Xu J et al (2010) Genetic reduction of striatal-enriched tyrosine phosphatase (STEP) reverses cognitive and cellular deficits in an Alzheimer's disease mouse model. Proc Natl Acad Sci U S A 107:19014–19019
10. Xu J, Chatterjee M, Baguley TD et al (2014) Inhibitor of the tyrosine phosphatase STEP reverses cognitive deficits in a mouse model of Alzheimer's disease. PLoS Biol 12:e1001923
11. Kulikova EA, Khotskin NV, Illarionova NB et al (2018) Inhibitor of striatal-enriched protein tyrosine phosphatase, 8-(Trifluoromethyl)-1,2,3,4,5-Benzopentathiepin-6-amine hydrochloride (TC-2153), produces antidepressant-like effect and decreases functional activity and protein level of 5-HT2A receptor in the brain. Neuroscience 394:220–231
12. Siemsen BM, Lombroso PJ, McGinty JF (2018) Intra-prelimbic cortical inhibition of striatal-enriched tyrosine phosphatase suppresses cocaine seeking in rats. Addict Biol 23: 219–229
13. Chatterjee M, Kwon J, Benedict J et al (2021) STEP inhibition prevents Abeta-mediated damage in dendritic complexity and spine density in Alzheimer's disease. Exp Brain Res 239: 881–890
14. Lee ZF, Huang TH, Chen SP et al (2021) Altered nociception in Alzheimer disease is associated with striatal-enriched protein tyrosine phosphatase signaling. Pain 162:1669–1680
15. Chatterji T, Gates KS (1998) DNA cleavage by 7-methylbenzopentathiepin: a simple analog of the antitumor antibiotic varacin. Bioorg Med Chem Lett 8:535–538
16. Lee AH, Chan AS, Li T (2002) Acid-accelerated DNA-cleaving activities of antitumor antibiotic varacin. Chem Commun:2112–2113
17. Lee AH, Chen J, Liu D et al (2002) Acid-promoted DNA-cleaving activities and total synthesis of varacin C. J Am Chem Soc 124: 13972–13973
18. Greer A (2001) On the origin of cytotoxicity of the natural product varacin. A novel example of a pentathiepin reaction that provides evidence for a triatomic sulfur intermediate. J Am Chem Soc 123:10379–10386
19. National Center for Biotechnology Information. PubChem BioAssay Database; AID=588619. https://pubchem.ncbi.nlm.

nih.gov/bioassay/588619. Accessed 3 Feb 2019

20. Suzuki Masaki [JP], Kondo Kazumi [JP], Kurimura Muneaki [JP], Valluru Krishna Reddy [In], Takahahi Akira [JP], Kuroda Takeshi [JP], Takahashi Haruka [JP], Fukushima Tae [JP], Miyamura Shin [JP], Ghosh Indranath [US], Dogra Abhishek [US], Harriman Geraldine [US], Elder Amy [US], Shimiza Satoshi [JP], Hodgetts Kevin J [US], Newcom Jason S [US]. Quinazolines as therapeutic compounds and related methods of use. (2014) US2014315886 (A1)
21. Suzuki Masaki [JP], Kondo Kazumi [JP], Kurimura Muneaki [JP], Valluru Krishna Reddy [In], Takahashi Akira [JP], Kuroda Takeshi [JP], Takahashi Haruka [JP], Fukushima Tae [JP], Miyamura Shin [JP], Ghosh Indranath [US], Dogra Abhishek [US], Harriman Geraldine [US], Elder Amy [US], Shimizu Satoshi [JP], Hodgetts Kevin J [US], Newcom Jason S [US]. Therapeutic compounds and related methods of use. (2015) US2015307477 (A1)
22. Witten MR, Wissler L, Snow M et al (2017) X-ray characterization and structure-based optimization of striatal-enriched protein tyrosine phosphatase inhibitors. J Med Chem 60: 9299–9319
23. Molina DM, Jafari R, Ignatushchenko M et al (2013) Monitoring drug target engagement in cells and tissues using the cellular thermal shift assay. Science 341:84–87
24. Prabhu N, Dai L, Nordlund P (2020) CETSA in integrated proteomics studies of cellular processes. Curr Opin Chem Biol 54:54–62
25. Martinez Molina D, Jafari R, Ignatushchenko M et al (2013) Monitoring drug target engagement in cells and tissues using the cellular thermal shift assay. Science 341:84–87
26. Dart ML, Machleidt T, Jost E et al (2018) Homogeneous assay for target engagement utilizing bioluminescent thermal shift. ACS Med Chem Lett 9:546–551
27. Martinez NJ, Asawa RR, Cyr MG et al (2018) A widely-applicable high-throughput cellular thermal shift assay (CETSA) using split Nano luciferase. Sci Rep 8:1–16
28. Romero C, Lambert LJ, Sheffler DJ et al (2020) A cellular target engagement assay for the characterization of SHP2 (PTPN11) phosphatase inhibitors. J Biol Chem 295:2601–2613
29. Lambert LJ, Romero C, Sheffler DJ et al (2020) Assessing cellular target engagement by SHP2 (PTPN11) phosphatase inhibitors. J Vis Exp
30. McNulty DE, Bonnette WG, Qi H et al (2018) A high-throughput dose-response cellular thermal shift assay for rapid screening of drug target engagement in living cells, exemplified using SMYD3 and IDO1. SLAS Discov 23:34–46
31. Henderson MJ, Holbert MA, Simeonov A et al (2020) High-throughput cellular thermal shift assays in research and drug discovery. SLAS DISCOVERY: Advancing the Science of Drug Discovery 25:137–147

Chapter 13

Target Deconvolution by Limited Proteolysis Coupled to Mass Spectrometry

Viviane Reber and Matthias Gstaiger

Abstract

Limited proteolysis coupled to mass spectrometry (LiP-MS) is a recent proteomics technique that allows structure-based target engagement profiling on a proteome-wide level. To achieve this, native lysates are first incubated with a compound, followed by a short incubation with a nonspecific protease. Binding of a compound can change accessibility at the binding site or induce other structural changes in the target. This leads to treatment-specific proteolytic fingerprints upon limited proteolysis, which can be analyzed by standard bottom-up MS-based proteomics. Here, we describe a basic LiP-MS protocol using the natural product rapamycin as an example compound. Along with the provided LiP-MS reference data available via ProteomeXchange with identifier PXD035183, this enables the straightforward implementation of the method by scientists with a basic biochemistry and mass spectrometry background. We describe how the procedure can easily be adapted to other protein samples and small molecules.

Key words Limited proteolysis, Structural proteomics, Target deconvolution, Mass spectrometry, Structural biology, Small molecules, Drug targets, Ligand–protein interaction, Target selectivity

1 Introduction

Target selectivity profiling in complex cellular proteomes represents a key step in the development of high-quality chemical probes. Mass spectrometry (MS)-based proteomics offers several essential approaches to analyze target selectivity of small molecules. This includes chemoproteomics, a selectivity profiling method on the basis of immobilized compounds for target affinity purification, and thermal protein profiling (TPP) or cellular thermal shift assay (CETSA), an orthogonal method for profiling compound-induced changes in the thermal stability of proteins [1]. Here, we describe a third method, namely limited proteolysis coupled to mass spectrometry (LiP-MS), a structural proteomics method that can detect changes in target protein structure caused by small-molecule bind-

Daniel Merk and Apirat Chaikuad (eds.), *Chemogenomics: Methods and Protocols*, Methods in Molecular Biology, vol. 2706,
https://doi.org/10.1007/978-1-0716-3397-7_13,

ing in complex protein lysates [2]. These approaches complement each other, and their respective advantages and challenges have been discussed previously [3].

LiP-MS relies on limited proteolysis using the sequence-unspecific protease proteinase K (PK) which has been shown to preferentially cleave flexible and accessible regions [4]. The binding of small molecules to proteins leads to treatment-dependent differences in flexibility and accessibility of the binding site and beyond, which in turn is captured by the change in cleavage sites of PK during the LiP step. Thus, the change in abundance of peptides generated by PK in the different treatment conditions can be used to identify regions where structural changes occur, providing a quantitative mass spectrometry-based readout for protein structural changes. Compared to the other MS-based target profiling methods, LiP-MS is the only method that allows both proteome-wide target deconvolution as well as identification of structural changes in target proteins upon compound binding. However, several factors that affect LiP-MS target deconvolution should be taken into account: (i) Detection of peptides near the compound binding sites requires good sequence coverage, and thus, LiP-MS is biased toward abundant proteins of a proteome. (ii) Not all compounds may affect PK accessibility to the target. (iii) Besides affecting the structure at the binding site, small molecules may also induce allosteric changes beyond this site. Since both structural alterations can affect PK cleavage patterns, it can be challenging to distinguish between the binding site and other structural changes based on LiP-MS results alone. For comprehensive profiling of the target space and analysis of target structural changes, LiP-MS should be complemented by orthogonal methods [5].

The following protocol represented in Fig. 1 is used for target engagement analysis in complex lysates. However, it can easily be adapted for LiP-MS of low-complexity samples, including purified proteins. First, proteins must be natively extracted from cells or tissue samples to ensure intact protein structures. The drug is added to the lysate followed by a brief incubation, after which the precisely timed limited proteolysis step is carried out. The exact timing and ratio of proteases to total protein are crucial to ensure reproducibility in the PK cleavage patterns. The generated protein fragments are further processed by reduction and alkylation of cysteines and a tryptic digest for subsequent MS analysis. Note that the peptides generated by the tryptic digest still reflect the fragments generated by PK during limited proteolysis. The peptides are measured using SWATH-MS, a variant of data-independent acquisition (DIA) as this yields more complete data and increased quantitative accuracy compared to data-dependent acquisition (DDA) methods [6]. In DDA methods, the most intense ions are selected for further fragmentation, whereas SWATH-MS can measure all ionized

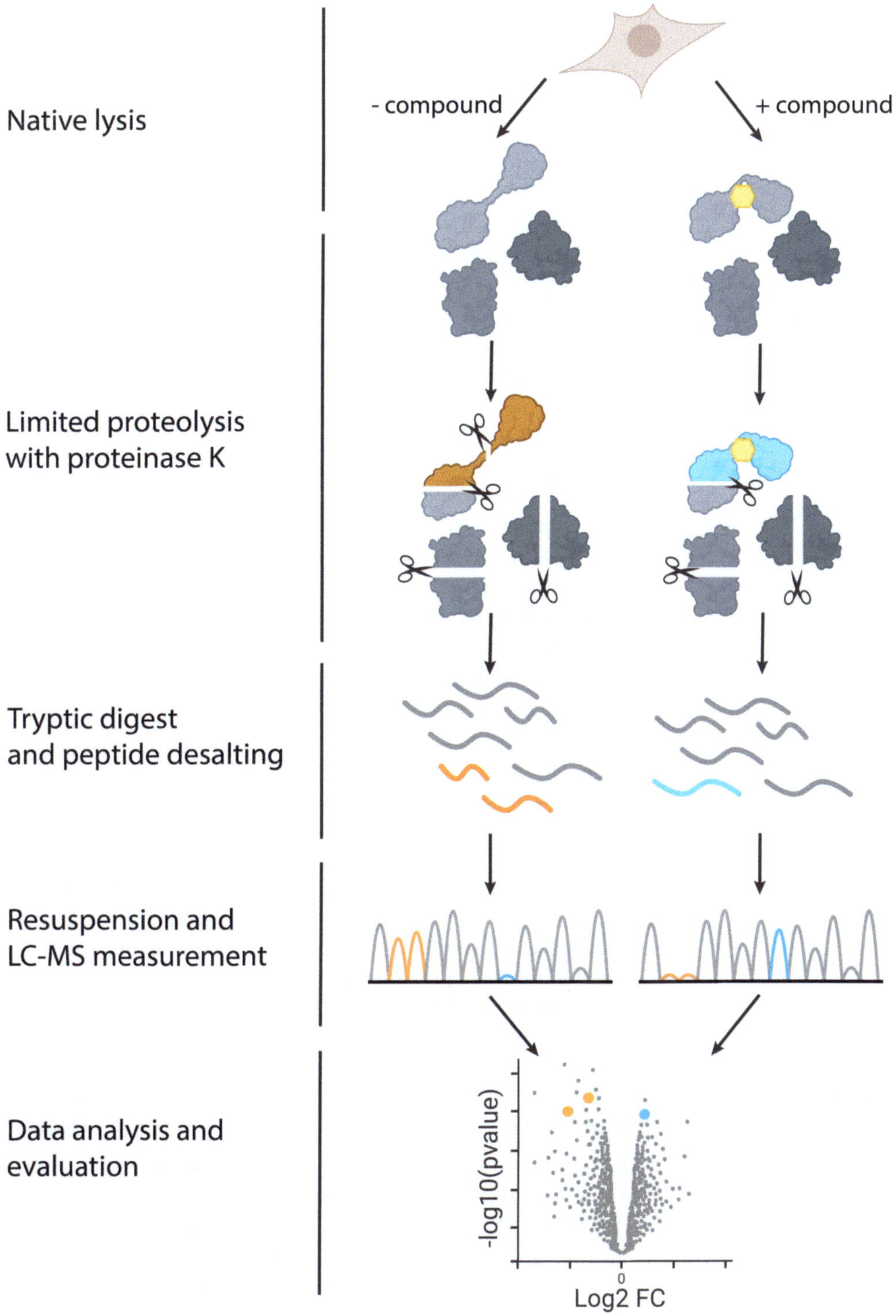

Fig. 1 A schematic representation of the LiP-MS workflow. The cells are lysed under native conditions to preserve the structures of the proteins. The lysate is treated either with a drug or with the vehicle as a mock treatment. Then, PK is added for exactly 5 min in the limited proteolysis step. This generates structure-specific protein fragments based on the treatment. The protein fragments are processed further with a tryptic digest and a desalting step to generate peptides for LC-MS measurement. The change in peptide abundance is quantified using MS to identify protein regions where structural changes occur to identify the binding sites of the added drug

peptides of a sample by fragmentation of precursor ions sequentially selected from larger isolation windows. Robust extraction of peptide level information from SWATH-MS spectra can be performed by applying prior knowledge obtained from peptide libraries generated from DDA measurements performed in parallel, via direct DIA approaches, or a combination of both using commercial software such as Spectonaut [7] or public software packages. The resulting peptide intensity data are further evaluated and visualized with the R package protti [8]. The fold change of the measured peptide intensities is calculated between the treatment and control conditions, and the significance of the results is assessed using a two-tailed moderated t-test, though other statistical tests are also commonly used. Due to the high number of significance tests calculated, multiple testing correction has to be done, for example using the method of Benjamini–Hochberg [9]. These calculations, among others, are implemented in protti. To facilitate the implementation of analysis and evaluation of LiP-MS data, we have deposited acquired data from a representative LiP-MS experiment using rapamycin as shown in Fig. 2 to the ProteomeXchange Consortium via the PRIDE [10] partner repository with the dataset identifier PXD035183. We recommend downloading this dataset to familiarize yourself with the data analysis pipeline before conducting your experiment.

2 Materials

Prepare all solutions fresh on the day of the experiment. We use ultrapure water with a resistance of 18 MΩ-cm at 25 °C and analytical grade reagents up to the desalting step, and then proceed with HPLC grade reagents. All the amounts specified are calculated for an experiment with two conditions (drug treatment and control) with four replicates per condition, resulting in eight samples.

2.1 Lysis

1. Cell pellet: snap frozen pellet from approximately 10^7 tissue culture cells to yield at least 800 μg of protein (*see* **Note 1**).
2. LiP buffer: 1 mM $MgCl_2$, 150 mM KCl, and 100 mM 4-(2-hydroxyethyl)-1-piperazineethanesulfonic acid (HEPES). We use 1 M stock solutions that we store for several months for each of the components. The 1 M HEPES stock solution should be adjusted to a pH of 7.4 using potassium hydroxide before use. The buffer should be stored on ice.
3. Water: 20 ml in 50-ml tube.
4. 70% ethanol: 20 ml in 50-ml tube.
5. 15-ml tube.
6. Snap cap tubes.
7. Centrifuge for snap cap tubes.

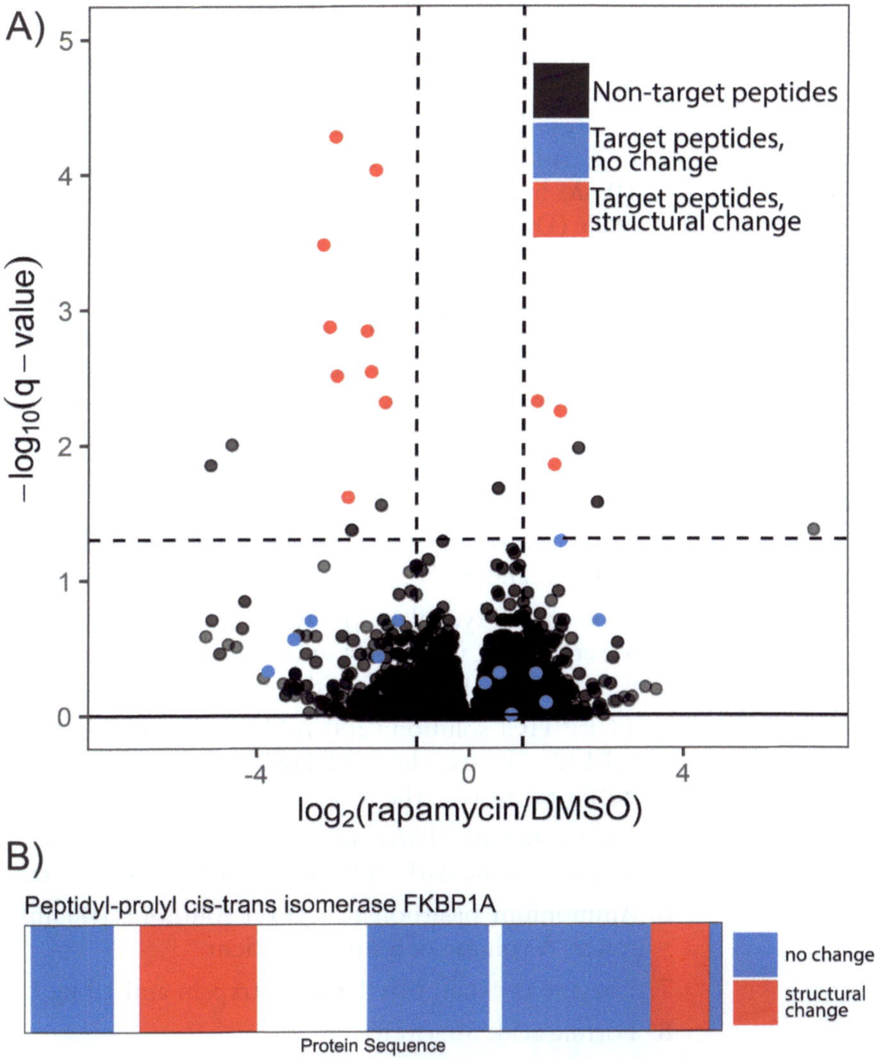

Fig. 2 (**a**) A volcano plot with fold changes in peptide intensities upon rapamycin treatment compared to DMSO control (x-axis: log2(fold change); y-axis: -log10(q-value)) as calculated using the proDA method implemented in protti. The blue dots represent peptides that map to the known rapamycin target FKBP1A which do not change upon rapamycin, whereas red dots indicate FKBP1A peptides which significantly change in the presence of rapamycin. Black dots indicate all other peptides quantified. Overall, 88,608 peptide precursors from 5317 proteins have been analyzed in the shown volcano plot. (**b**) A barcode plot that shows where structural changes are observed in FKBP1A along the primary sequence of the protein (N- to C-term). Note that the rapamycin induced structural changes yield changes in the abundance of many overlapping peptides (shown in Fig. **a**) from two regions of the protein indicated in red

2.2 Limited Proteolysis Step

1. Thermal cycler: Set to 25 °C in one chamber and 99 °C in another chamber.
2. Bucket of ice.
3. Workspace next to the thermal cycler: a multichannel pipette for volumes of 1 μl and 5 μl, appropriate pipette tips, a waste bin for the tips, and a timer.

4. Empty strip of eight 0.2-ml tubes with lids.
5. Strip of eight 0.2-ml tubes containing the drug or vehicle: The necessary volume is only 1 μl per tube but supplying 5 μl ensures that there is enough volume to allow easy pipetting. Add 5 μl of the drug (e.g., 510 μM rapamycin for a 10 μM final concentration) or the vehicle (here dimethyl sulfoxide (DMSO)) in four tubes each (*see* **Note 2**). Keep the tubes at room temperature.
6. Strip of eight 0.2-ml tubes containing PK: The necessary volume is only 5 μl per tube but supplying 10 μl ensures that there is enough volume to allow easy pipetting. Add 10 μl PK at concentration of 0.2 μg/μl to each tube ensure an enzyme: substrate ratio of 1:100. Store the tubes on ice.

2.3 Tryptic Digest

1. Snap cap tubes.
2. Thermomixer capable of shaking the snap cap tubes at 800 rpm at 30 °C and 37 °C with a lid.
3. Sodium deoxycholate (DOC) solution: 10% (w/v) DOC in water. Vortex vigorously until the DOC is fully dissolved and wait until the foam has settled. A volume of 1 ml is sufficient.
4. TCEP-HCl solution: 200 mM TCEP-HCl dissolved in 1 M HEPES. We use the 1 M HEPES stock solution for the LiP buffer to dissolve the reagent. A volume of 50 μl is sufficient.
5. Iodoacetamide (IAA) solution: 1 M IAA in water. Store the solution in the dark. A volume of 100 μl is sufficient.
6. Ammonium bicarbonate (ABC) solution: 100 mM ABC in water. A volume of 5 ml is sufficient.
7. Enzyme cocktail: Mix 10 μg of trypsin and 10 μg Lys-C.
8. Formic acid: 50% in water.
9. 0.2-μm PVDF filter plate.
10. Collection plate: suitable for PVDF filter plate.
11. Vacuum manifold: suitable for PVDF filter plate.

2.4 Peptide Desalting

1. MS grade methanol.
2. Buffer A: 0.1% formic acid in HPLC grade water.
3. Buffer B: 50% HPLC grade acetonitrile and 0.1% formic acid in HPLC grade water.
4. C18 columns or plates: Ensure that the amount of C18 material corresponds to the amount of peptide in each sample.
5. Waste collection tubes or plates: suitable for C18 columns or plate.
6. Eluate collection tubes or plates: suitable for C18 columns or plate.

2.5 Peptide Resuspension

1. Thermomixer capable of shaking the snap cap tubes at 800 rpm at 25 °C.
2. Water bath sonicator.
3. Centrifuge for snap cap tubes.
4. Labeled MS vials.
5. Retention time peptide standard: Resuspend according to manufacturer's instructions. We use the iRT peptides by from Biognosys as this kit is compatible with Spectonaut, the search software we use. Other peptide standards can be used, but need to be taken into account during the search step. Using a retention time peptide standard is optional.

2.6 MS Measurement

1. LC-MS setup: We use an Orbitrap Fusion Lumos Tribrid mass spectrometer equipped with an Acquity M-Class UPLC System. Comparable setups are also sufficient.
2. LC method: To separate the peptides on a column packed with 1.9 μm C18 beads, use a linear gradient of 3% to 35% acetonitrile in water with 0.1% FA over the course of 120 min at flow rate of 300 nl/min.
3. MS/MS method: We suggest using a DIA method, although DDA measurements also provide data of high quality. The MS1 spectra are acquired in a scan range of 350–1500 m/z with an Orbitrap resolution of 120,000 with the normalized AGC target set to 200% with a maximum injection time of 264 ms and the RF lens set to 50%. The targeted MS2 spectra are acquired for the desired masses with 20 variable isolation windows and fragmented with a HCD of 30%. The spectra are measured with an Orbitrap resolution of 30,000 with variable scan ranges listed in Table 1 and the AGC target set to 200% with a maximum injection time of 54 ms and the RF lens set to 50%. Other instrument setups may require different settings. Either the manufacturer or an experienced MS user can provide details for the optimal settings for your instrument.

3 Methods

3.1 Lysis

1. Thaw the cell pellet on ice. Add 200 μl LiP buffer before the pellet is thawed completely. Avoid adding more buffer as the protein concentration of the final lysate should be above 2 μg/μl. The protein concentration will be determined and adjusted appropriately in a later step.
2. Gently pipette the pellet several times to homogenize the sample and transfer it to a 15-ml tube (*see* **Note 3**).
3. Clean the pestle before use in 70% ethanol and water.

Table 1
Listed are the windows used in the DIA method for the MS/MS measurement. The window sizes have been optimized for complex peptide samples

Window	Lower bound (m/z)	Upper bound (m/z)	Width (m/z)
1	349.5	381.5	32
2	380.5	408.5	28
3	407.5	432.5	25
4	431.5	458.5	27
5	457.5	482.5	25
6	481.5	506.5	25
7	505.5	530.5	25
8	529.5	553.5	24
9	552.5	578.5	26
10	577.5	605.5	28
11	604.5	633.5	29
12	632.5	662.5	30
13	661.5	693.5	32
14	692.5	727.5	35
15	726.5	766.5	40
16	765.5	809.5	44
17	808.5	862.5	54
18	861.5	929.5	68
19	928.5	1029.5	101
20	1028.5	1498.5	470

4. Use a Kimble pellet pestle with a cordless motor to lyse cells in the 15-ml tube using 10 pulses followed by 1 min on ice, repeated 10 times (*see* **Note 4**).
5. Clean the pestle after use in 70% ethanol and water.
6. To remove membranes and other debris, transfer the lysate to a snap cap tube and centrifuge at 20000× *g* for 15 min at 4 °C. Collect the supernatant in a new tube.
7. Determine the total protein concentration using a Pierce BCA protein assay kit according to manufacturer's instructions. Dilute the lysate to 2 μg/μl in LiP buffer and distribute 50 μl into each 0.2-ml tube on a strip for easy handling. The total

protein amount is 100 μg per sample (*see* **Note 5**). Prepare four replicates per condition for a total of eight samples.

3.2 Limited Proteolysis Step

1. Use the multichannel pipette to add 1 μl of the prepared drug solution or vehicle to the lysate and mix. Mix by pipetting gently to avoid bubbles.
2. Incubate at 25 °C for 5 min. Start the timer exactly when the drug is added to the lysate (*see* **Note 6**).
3. Change the pipetting volume of the pipette used from 1 μl to 5 μl, or switch pipette.
4. Using the multichannel pipette to add 5 μl of PK to the lysate and mix as during the drug treatment.
5. Incubate at 25 °C for exactly 5 min. Start the timer exactly when PK is added to the lysate.
6. Transfer the tubes to 99 °C to inactivate PK (*see* **Note 7**).
7. Transfer the tubes to ice or to 4 °C to cool for at least 5 min.

3.3 Tryptic Digest

1. Add the same volume of 10% DOC for a final concentration of 5% DOC (56 μl 10% DOC in 56 μl sample). Mix well by gently pipetting until the solution is clear, and transfer the sample to fresh snap cap tubes. At this point, the samples can be frozen until further processing (*see* **Note 8**).
2. Reduce disulfide bonds with a final concentration of 5 mM TCEP (2.9 μl 200 mM TCEP added to 112 μl sample). Incubate the samples for 40 min at 37 °C and 800 rpm on a thermomixer.
3. Alkylate samples with a final concentration of 40 mM IAA (4.8 μl 1 M IAA added to 114.9 μl sample). Incubate for 30 min at 30 °C in the dark, at 800 rpm.
4. Dilute sample with 100 mM ABC to a DOC concentration of 1% (1:5) (478.7 μl 100 mM ABC added to 119.7 μl sample).
5. Add 1 μg Lys-C and 1 μg trypsin from the pre-mixed stock to each sample. Incubate overnight at 37 °C and 800 rpm.
6. Stop the digestion by adding 50% formic acid to reach a final concentration of 2% (1:25) (approx. 25 μl 50% formic acid added to 598.4 μl). Shake for 5 min at 37 °C at 800 rpm.
7. Remove precipitated DOC by filtration in a fume hood using a 96-well vacuum manifold and 0.2 μm PVDF filter plates.

3.4 Peptide Desalting

1. Wash C18 columns with 1 volume methanol, followed by 1 volume buffer B. The manufacturer specifies the volume for the various C18 columns. Spin down after each washing step.

The manufacturer specifies the centrifugation settings. Appropriately discard the waste.

2. Equilibrate the columns twice with 1 volume buffer A. Spin down after each step.
3. Transfer sample onto column and collect the flow-through. This might require multiple steps if the volume of the column is smaller than the total sample volume. Spin down after each step.
4. Wash three times with 1 volume buffer A. Spin down after each step.
5. Elute twice with 100 μl buffer B for a final volume 200 μl.
6. Dry samples in a vacuum concentrator at 45 °C for 2–3 h until fully dry.
7. Store samples at −20 °C until resuspension.

3.5 Peptide Resuspension

1. Resuspend peptides in 50 μl buffer A to a final concentration of 1 μg/μl. We assume a 50% loss when starting out with 100 μg protein. Rinse walls of the tube with buffer A to solubilize as much of the peptides as possible.
2. Shake the samples at 25 °C for 5 min at 800 rpm.
3. Sonicate the samples for 5 min in a water bath sonicator at 25 °C.
4. Spin the samples at 20000× *g* for 5 min at 15 °C.
5. Transfer 9 μl of the supernatant of the samples to MS sample vials and optionally add 1 μl of the retention time peptide standard.
6. In the short term, samples can be stored at 4 °C until measurement. If the samples must be stored for more than a day, we recommend storing them at −20 °C.

3.6 MS Measurement

1. Shake out bubbles from the samples before measuring.
2. Measure 2% (1 μl injection volume) of each sample in a randomized order according to the methods described above. We expect base peak intensities, also called normalization levels (NL) of 1e10 for our MS setup. This might be different for other sample types and instruments. If the measured total intensity is higher or lower than expected, the amount injected can be adjusted.
3. After measurement, replace the caps of the vials with new ones to store the samples at −20 °C.

3.7 Data Evaluation

1. We use Spectronaut version 15 to identify and quantify peptides from the raw DIA data. We apply the *directDIA* feature

and use the factory settings except that we expand the search space to include semi-specific peptides and that the differential abundance grouping is set to precursors and not to proteins. This allows us to interpret the data on a peptide level. Other software can be used, though results may vary based on the specific algorithms implemented in the various programs.

2. The peptide intensity data are exported from Spectronaut and further analyzed in Rstudio using R 4.0.2. We use the R package protti for data normalization, filtering, quality control, differential abundance calculation, and significance testing.
3. To check the quality of the acquired MS data, we use various quality controls. The intensity distributions and peptide identifications should be comparable across samples. The proportion of semi-specific peptides should be between 10% and 40% and should be comparable across samples. Principal component analysis (PCA) can be used to assess how the samples cluster. Lastly, we check the coefficients of variation (CVs) of each condition. We expect the median CV to be below 20% for lysates but may be higher for purified samples (*see* **Note 9**).
4. The final readout of the assay is whether the measured peptides significantly change between conditions. This is defined via a significance cutoff of *adjusted p-val* < *0.05* and *differential abundance* > *2*. This can be calculated in protti and represented easily in a volcano plot as demonstrated in Fig. 2a. The significant peptides represent regions of structural change due to the drug treatment.
5. In order to better understand the change in peptides, they can be mapped to the primary structure of the protein as shown in Fig. 2b. This can be compared to known domain and active site annotations to generate hypotheses on the nature of the structural changes.
6. If available, the structural changes can be mapped to crystal structures of the protein to better understand how the structural changes relate to each other in the 3D structure. Alternatively, structural predictions can be used. This is also implemented in protti.

4 Notes

1. Typically, a HEK293 cell pellet from a 10 cm cell culture dish at 80% confluence is sufficient. Pellets from tissues or other organisms such as *E. coli* and yeast can also be used. Purified proteins can also be used and are of advantage if peptide coverage in lysate samples is not sufficient for the protein of interest. In this case, the enzyme:substrate ratio must be kept at

1:100 for PK, trypsin, and Lys-C, which requires adjustment of all enzyme concentrations. The volumes and concentrations of the other reagents remain as specified.

2. The drug concentration to be used will vary based on the sample and the drug. Typically, we add between 1 μM and 10 μM of the drug to the lysate. The structural changes cannot be observed if the drug concentration is too low, and false-positive structural changes can be observed if too much of the drug is added. It is also possible to measure multiple drug concentrations and measure the change in peptide intensity in response to the drug concentration. Data analysis pipelines for multiple drug concentrations are also available in protti.

3. In order to make sure that the pipetting of the lysate is gentle to avoid protein denaturation and complex dissociation, choose the largest convenient pipette tip. Additionally, we suggest cutting the pipette tip a few millimeters shorter to increase the size of the opening in order to transfer the lysate.

4. Make sure to avoid warming and bubbles during the lysis process. Avoid warming by holding the top of the tube. Avoid bubbles by angling the motorized pestle appropriately.

5. If you do not have enough material, you can dilute the lysate to 1 μg/μl for a final protein amount of 50 μg. This requires adjustment of the PK, trypsin, and Lys-C concentrations as described in **Note 1**, but the volumes otherwise stay the same.

6. It is crucial to get the timing of the steps right to ensure the reproducibility of the experiments. We recommend starting the timer the second that the drug is added to the sample. This is done by pipetting into tubes held in a rack with one hand while holding the timer with the other hand. It is easiest to count up and start the PK incubation precisely at minute 5, start the inactivation at minute 10, and transfer to ice at minute 15. In case one step starts late, make note of this delay in order to ensure that all subsequent steps are still precisely 5 min (e.g., if the PK incubation starts at 5:06, start the inactivation at 10: 06). In order to get the timing right, take the tubes out of the thermal cycler 20–30 seconds before pipetting to open the tubes. It takes some time to draw up the reagent, check that all tips of the multichannel pipette are filled correctly, and then, dispense into the sample precisely 5 min after the previous step. Closing the tubes, mixing, and centrifugation should be done quickly but with care, and count as part of the incubation time.

7. During the incubation at 99 °C, the caps of the tubes may pop off. It is therefore important to open the thermal cycler carefully and hold the caps in place.

8. After adding DOC, samples can be frozen. Thaw the samples at room temperature. If any DOC has precipitated, shake the samples at 60 °C for 10 min at 800 rpm. Wait for the samples

to cool to room temperature before continuing with the protocol.

9. The quality controls measures are used to ensure that the data are good enough to use them to identify structural changes. If individual samples are of low quality, these can be remeasured. If the quality does not improve upon remeasurement, the sample can be removed. If multiple samples in the data are of insufficient quality, we recommend repeating the experiment. Note that poor clustering in a PCA plot is not sufficient to exclude a sample, as very few structural changes are expected and thus the samples often overlap between conditions. We recommend that a sample is excluded if the median intensity or the peptide identifications are off by more than 30% compared to all other samples or if the proportion of semi-tryptic peptides is more than 5% different compared to the other samples.

Acknowledgements

We would like to acknowledge valuable inputs from Ludovic Gillet and the rest of the Picotti lab for optimising this protocol. This work was supported by the Innovative Medicines Initiative 2 Joint Undertaking (JU) under grant agreement No 875510. The JU receives support from the European Union's Horizon 2020 research and innovation program and EFPIA and Ontario Institute for Cancer Research, Royal Institution for the Advancement of Learning McGill University, Kungliga Tekniska Hoegskolan, Diamond Light Source Limited.

References

1. Molina DM, Jafari R, Ignatushchenko M et al (2013) Monitoring drug target engagement in cells and tissues using the cellular thermal shift assay. Science 341(6141):84–87. https://doi.org/10.1126/science.1233606
2. Piazza I, Beaton N, Bruderer R et al (2020) A machine learning-based chemoproteomic approach to identify drug targets and binding sites in complex proteomes. Nat Commun 11(1):4200. https://doi.org/10.1038/s41467-020-18071-x
3. Meissner F, Geddes-McAlister J, Mann M, Bantscheff M (2022) The emerging role of mass spectrometry-based proteomics in drug discovery. Nat Rev Drug Discov 21(9): 637–654. https://doi.org/10.1038/s41573-022-00409-3
4. Feng Y, De Franceschi G, Kahraman A et al (2014) Global analysis of protein structural changes in complex proteomes. Nat Biotechnol 32(10):1036–1044. https://doi.org/10.1038/nbt.2999
5. Hendricks JA, Beaton N, Chernobrovkin A et al (2022) Mechanistic insights into a CDK9 inhibitor via orthogonal proteomics methods. ACS Chem Biol 17(1):54–67. https://doi.org/10.1021/acschembio.1c00488
6. Gillet LC, Navarro P, Tate S et al (2012) Targeted data extraction of the MS/MS spectra generated by data-independent acquisition: a new concept for consistent and accurate proteome analysis. Mol Cell Proteomics 11(6): O111.016717. https://doi.org/10.1074/mcp.O111.016717
7. Bruderer R, Bernhardt OM, Gandhi T et al (2015) Extending the limits of quantitative proteome profiling with data-independent acquisition and application to acetaminophen-treated three-dimensional liver microtissues. Mol Cell Proteomics 14(5):1400–1410.

https://doi.org/10.1074/mcp.M114.044305

8. Quast J-P, Schuster D, Picotti P (2022) Protti: an R package for comprehensive data analysis of peptide- and protein-centric bottom-up proteomics data. Bioinforma Adv 2(1):vbab041. https://doi.org/10.1093/BIOADV/VBAB041
9. Benjamini Y, Hochberg Y (1995) Controlling the false discovery rate: a practical and powerful approach to multiple testing. J R Stat Soc Ser B 57(1):289–300. https://doi.org/10.1111/j.2517-6161.1995.tb02031.x
10. Perez-Riverol Y, Bai J, Bandla C et al (2022) The PRIDE database resources in 2022: a hub for mass spectrometry-based proteomics evidences. Nucleic Acids Res 50(D1):D543–D552. https://doi.org/10.1093/nar/gkab1038

Chapter 14

Global Assessment of Drug Target Engagement and Selectivity of Covalent Cysteine-Reactive Inhibitors Using Alkyne-Functionalized Probes

Elisabeth M. Rothweiler and Kilian V. M. Huber

Abstract

Covalent inhibitors are emerging as a promising therapeutic means for efficient and sustained targeting of key disease-driving proteins. As for classic non-covalent inhibitors, understanding target engagement and selectivity is essential for determining optimal dosing and limiting potential on- or off-target toxicity. Here, we present a complementary activity-based protein profiling (ABPP) strategy for unbiased proteome-wide profiling of cysteine-reactive inhibitors based on two orthogonal approaches. We illustrate the use of clickable alkyne probes for in-gel fluorescence and mass spectrometry studies using a series of therapeutic XPO1 inhibitors as an example.

Key words Activity-based protein profiling, Covalent inhibitors, Cysteine, Target engagement, Chemoproteomics

1 Introduction

The development of new medicines requires demonstrating the safety, efficacy, and quality of novel therapeutics. This comprises a thorough grasp of the drug candidate's physicochemical qualities and its pharmacological effects, both on-target and off-target, which may result in undesired side effects [1–3]. Due to their supposed inherent hyperreactivity and promiscuity, electrophilic covalent warheads have generally been considered as problematic from a chemistry perspective. There are currently over 30 FDA-approved drugs whose mechanism of action involves the formation of a covalent bond with their respective protein target (s) [4] and several clinical candidates in development feature reactive chloroacetamide and acrylamide moieties [5]. These advances have been enabled by recent studies indicating that such cysteine-reactive inhibitors can be optimized in terms of reactivity and selectivity [6, 7]. Nonetheless, extensive profiling of covalent

Daniel Merk and Apirat Chaikuad (eds.), *Chemogenomics: Methods and Protocols*, Methods in Molecular Biology, vol. 2706, https://doi.org/10.1007/978-1-0716-3397-7_14,

drug candidates remains an important task to confirm target engagement and specificity across tissues. Unlike biochemical methods that rely on purified recombinant proteins, chemical biology methods such as chemoproteomics survey drug–target interactions directly in cells or lysate. Importantly, these approaches consider intracellular effects such as different posttranslational modifications, metabolites, co-factors, or complex formation, all of which may affect drug-target binding. Various disease-relevant model systems and even patient material can be examined [8]. Studying drug target occupancy in such circumstances, in particular, can help to identify appropriate dosage regimens [9]. Using the covalent exportin 1 (XPO1) inhibitors selinexor and eltanexor [10] as examples, we show how an ABPP chemoproteomic workflow can be utilized to assess inhibitor selectivity and target engagement across the proteome. ABPP can be performed with both in-gel fluorescence and mass spectrometry (MS) readouts and represents a powerful means to query the on- and off-targets of covalent compounds [11] (Fig. 1).

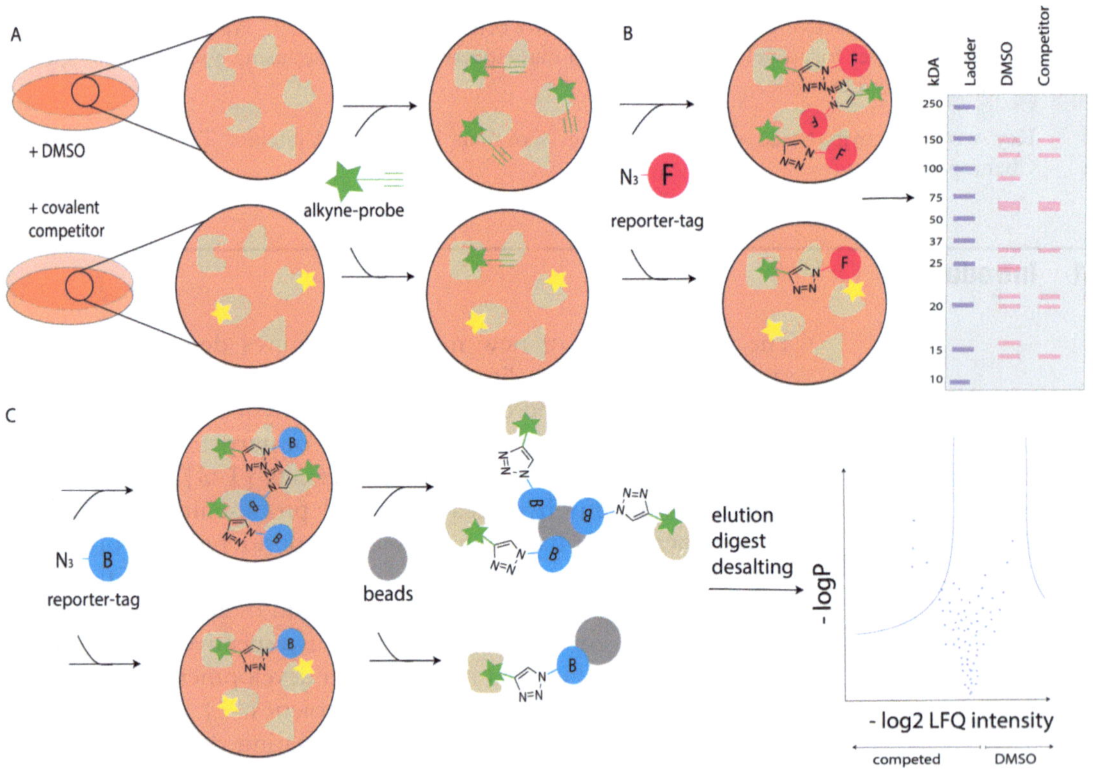

Fig. 1 ABPP experimental workflow. (**a**) Cells or lysate are pre-treated with DMSO or competitor (drug or compound of interest) before addition of the alkyne-functionalized affinity probe. (**b**) Subsequent addition of a fluorescent reporter tag allows for visualization of protein bands via SDS-PAGE. (**c**) Alternatively, a biotin tag is introduced for subsequent affinity enrichment on streptavidin-coated beads. After elution, digestion, and desalting, samples are analyzed via LC-MS/MS followed by bioinformatic analysis

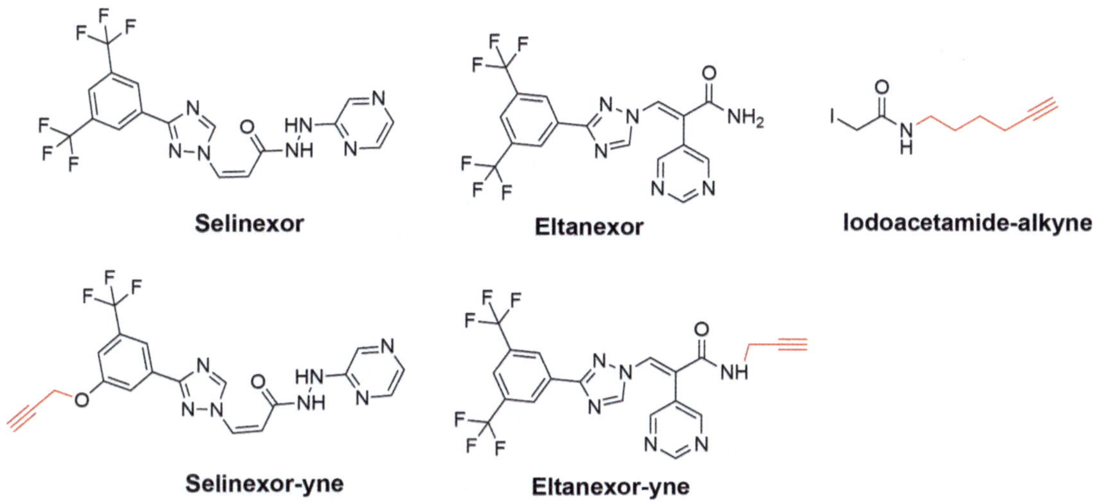

Fig. 2 Structures of the XPO1 inhibitors selinexor and eltanexor and respective affinity probes

Iodoacetamide (IA) is a broadly cysteine-reactive compound that has been used extensively to interrogate cysteine reactivity on a proteome-wide scale [12, 13]. By installing a terminal alkyne onto IA (Fig. 2), the resulting probe can be reacted with azide-containing fluorescent reporter molecules or affinity enrichment tags (e.g., biotin for streptavidin enrichment) in copper-catalyzed azide-alkyne cycloaddition reactions (CuAAC) [11]. Early examples of so-called activity-based probes were developed to measure serine reactivity; however, subsequent efforts have provided a broad range of probes that can survey lysine, methionine, or tyrosine residues [14–21]. Parallel treatment of cells or lysate with an unmodified competitor (e.g., the drug or compound of interest) allows for drug–target identification and establishment of dose–response curves. Alternatively, the drug or compound of interest can be chemically modified to append an alkyne tag for the enrichment of drug-binding proteins. This approach puts more emphasis on the biophysical interactions of the core chemical scaffold and requires knowledge of structure–activity relationships to inform the design of respective affinity probes ensuring their biological activity is not altered. Examples for affinity probe design and procedures for IAA- and compound-centric ABPP workflows for the clinical XPO1 inhibitors selinexor and eltanexor are provided below (Fig. 2).

2 Materials

Prepare solutions using MS-grade solvents at room temperature and use analytical-grade reagents. Prepare and store reagents at room temperature or as indicated by the supplier. Grow cells at 37 °C and 5% CO_2 atmosphere in a standing incubator, and use media as indicated by the supplier. Store harvested cell pellets at −80 °C.

2.1 Preparation of Buffers for Click Reaction and Cell Lysis

1. Alkyne compound: compound 100 mM stock in DMSO.
2. RIPA lysis buffer: 50 mM Tris–HCl, 150 mM NaCl, 1% NP-40, 0.5% sodium deoxycholate, and 0.1% SDS, pH 7.4. Weigh in 7.88 g of Tris–HCl and 8.77 g of NaCl, add 1000 water, and stir with a magnetic stirrer until the solids are dissolved. Add 100 mL of a 10% NP-40 solution, and add 100 mL of a 10% sodium deoxycholate solution and 1 mL of a 10% SDS solution. Add 1x ETDA-free complete protease inhibitor (Roche Diagnostics) and adjust pH.
3. TBTA stock: Tris(benzyltriazolylmethyl)amine 10 mM stock in DMSO. Weigh in 5.3 mg and dissolve in 1 mL DMSO.
4. $CuSO_4$ stock: 50 mM stock in water. Dissolve 7.98 mg $CuSO_4$ in 1 mL water.
5. TCEP stock: Tris(2-carboxyethyl)phosphine 50 mM stock in water. Dissolve 14.3 mg in 1 mL water.
6. Cy5.5 azide stock for in-gel fluorescence: 10 mM in DMSO. Dissolve 7.0 mg in 1 mL DMSO and aliquot to 20 μL each.
7. Azide-PEG-biotin stock for affinity enrichment: Azide-PEG3-biotin 10 mM stock. Dissolve 10 mg in 2.25 mL DMSO and aliquot to 10 μL.
8. Protein precipitation: 2 volumes of MeOH, 0.5 volumes of $CHCl_3$, and 1 volume of water.
9. Re-suspension buffer: 1x PBS, 2% SDS.

2.2 In-Gel Fluorescence

1. 4x Laemmli buffer: add freshly 20% DTT.
2. Pre-cast gels: 4–12% Criterion XT Bis–Tris Protein Gel (Bio-Rad).
3. Equipment: LI-COR Odyssey (LI-COR).

2.3 Affinity Enrichment, Digestion, and Desalting

1. Beads: Neutravidin agarose resin beads.
2. Wash buffer: 0.2% SDS in 1x PBS.
3. 2x Laemmli buffer: dilute 4x with PBS and add 20% DTT.
4. Equipment: heat block, rotating wheel, shaking incubator, and SpeedVac.
5. DTT stock: 200 mM in 0.1 M Tris.
6. Iodoacetamide: 200 mM in 0.1 M Tris.
7. Digestion buffer: 6 M urea in 0.4 M Tris. To prepare the Tris buffer, add 12.1 g Tris base to 200 mL water, adjust pH to 7.4, and fill up to 250 mL. Use 2 g urea and 1.25 mL 0.4 M Tris and fill up to 5 mL with water.
8. Trypsin: 1:50 ratio of trypsin to protein.

9. SOLA HRP SPE cartridge.
10. Buffer A: 65% MeCN and 0.1% formic acid.
11. Buffer B: 2% MeCN with 0.1% formic acid.

3 Methods

3.1 Reactive Probe Treatment and Cell Lysis

1. Grow the cells in indicated media to 80–90% confluency in a suitable dish or flask.
2. For in-gel fluorescence, use 100 μL sample volume and 100 μg protein sample (1 mg/mL), for MS-readout scale up to 500 μL and 1 mg for each condition (2 mg/mL).

   For competitive conditions, pre-treat the cells with the desired concentration of competitor, usually starting with 100 μM for 1 h at 37 °C. Add an equal amount of DMSO vehicle to the control batch (*see* **Note 1**).
3. Add the alkyne-modified probe at 100 μM to label reactive residues and incubate for 1 h at 37 °C (*see* **Note 2**).
4. Harvest the cells by pelleting and wash with 1x DBPS. Lyse cells in RIPA buffer containing 1x EDTA on ice for 30 min, followed by centrifugation at 14,000× *g*.
5. Snap-freeze the pellet in liquid nitrogen and store at −80 °C or proceed with step 6.
6. Adjust the protein concentration to 1 mg/mL using the DC protein assay kit 2 (Bio-Rad) and pipette 100 μL sample solution per 1.5 mL Eppendorf tube.

3.2 Click Reaction

1. Prepare the click master mix as a 1:2:1:2 mixture (volume of 6 μL per 100 μL sample volume) of TBTA, $CuSO_4$, Cy5.5 azide reporter tag, and TCEP and add immediately to the respective samples (*see* **Note 3**). For affinity enrichment, exchange the fluorescent tag for biotin probe, proceed to step 5, and then continue with procedure 3.4. Affinity enrichment. Incubate 1 h at RT shaking at 750 rpm in a shaking incubator.
2. Quench the reaction by adding 5 mM EDTA (500 mM stock, 100x dilution).

   For affinity enrichment and WB readout, retain 10% of the volume (10 μL) as "click"-control sample.
3. Precipitate the proteins using methanol–chloroform extraction: Add 2 sample volumes of MeOH, 0.5 volumes of $CHCl_3$ and 1 volume of water to the sample and vortex briefly (*see* **Note 4**).
4. Centrifuge the sample at 14,000× *g* for 2 min at RT and remove the supernatant (*see* **Note 5**).

5. Add 1000 μL MeOH and shake gently before centrifugation at 14,000× *g* for 5 min at RT. Remove the supernatant from the protein pellet and let dry for 5–10 min at RT.

3.3 In-Gel Fluorescence Detection

1. Re-suspend pellet in 2% SDS/PBS to 1 mg/mL. Take 10 μL and add 4 μL 4x Laemmli buffer, and boil for 5 min at 95 °C in a heat block.
2. Load 10 μL per pocket into 4–12% Bis–Tris gel in 1x MES running buffer.
3. Run 55 min 170 V and image using the Li-COR.
4. Perform a Coomassie stain of the gel to visualize protein loading.

3.4 Affinity Enrichment

1. Prepare the streptavidin beads (for 100 μg sample use 20 μL beads). Take 40 μL of a 50% slurry for 20 μL bead bed volume for each sample. Wash beads three times with 300 μL of 0.2% SDS in PBS. Agitate the beads in the wash solution, followed by centrifugation for 2 min at 2000× *g* at RT to re-pellet the beads. Remove the supernatant.
2. Re-suspend the sample in 10 μL 2% SDS in PBS and vortex. Dilute to 0.2% SDS by adding 90 μL PBS.
3. Add the samples to the beads and retain 10% (e.g., 10 μL of 100 μL) as input control for WB (*see* **Note 6**).
4. Incubate the samples for 1 h at RT spinning at 10 rpm on a rotating wheel.
5. Centrifuge and pellet the beads for 2 min at 2000× *g* at RT. Transfer supernatant to a fresh 1.5-mL LoBind Eppendorf tube and keep as supernatant control.
6. Wash the beads thoroughly three times with 300 μL of 0.2% SDS in PBS. Centrifuge for 2 min at 2000× *g* at RT and remove the supernatant.
7. Add 16 μL 2x Laemmli buffer (dilution of 4x Laemmli 1:2 in PBS) to the beats and boil in a heat block at 90 °C for 10 min. Let cool to RT and centrifuge the beads for 2 min at 2000× *g* at RT. Retain the supernatant as eluent. Use 12 μL of each click and supernatant control sample and add 4 μL of 4x Laemmli buffer and boil at 90 °C for 5 min. For WB, load 10 μL per sample and control with a suitable ladder.

3.5 MS Sample Preparation

1. Repeat the methanol–chloroform precipitation (Subheading 3.2, **steps 3–5**).
2. This time, re-suspend the sample in 50 μL 6 M urea in 0.4 M Tris.

3. Reduce samples by adding 40 mM DTT (200 mM stock in 0.1 M TRIS). Vortex and incubate for 60 min at RT.
4. Alkylate samples by adding 40 mM IA (200 mM stock in 0.1 M TRIS). Vortex and incubate for 30 min at RT while rotating at 10 rpm, covered in aluminum foil.
5. Dilute the sample to <1 M urea by adding 250 μL water.
6. For in-solution digestion, add 1:50 (trypsin:protein) and incubate overnight at 37 °C in a shaking incubator at 450 rpm.

3.6 Desalting

1. Acidify the sample with formic acid 1% to 0.1% (300 μL sample volume and 3 μL of 1% formic acid, corresponding to 1% of volume).
2. Place desalting columns on top of a vacuum and equilibrate columns with 0.5 mL solution B. Do not let the column bed run dry (*see* **Note 7**).
3. Add 1 mL solution A and let drip through, and then, load 300 μL sample onto the column by dropping directly in the middle of the column (*see* **Note 8**). Load sample by applying the vacuum. Add 1 mL solution A to wash the sample and apply vacuum.
4. Elute with 0.6 mL into fresh 2 mL LoBind tube by pushing through with a pipette tip and creating pressure with air (*see* **Note 9**).
5. Evaporate the solvent in the SpeedVac at 30 °C until the samples are dry.
6. Re-solve in 20 mL buffer B and transfer to mass spectrometry-grade vials.

3.7 MS Instrument Settings and Data Analysis

1. Digested samples are analyzed by nano-UPLC–MS/MS e.g., using a Dionex UltiMate 3000 Nano UPLC fitted with an EASY spray column (75 μm × 500 mm, 2 μm particle size, Thermo Scientific), coupled to an Orbitrap Q Exactive instrument. A 60 min gradient of 0.1% (v/v) formic acid in 5% (v/v) DMSO to 0.1% (v/v) formic acid with 35% (v/v) acetonitrile in 5% (v/v) DMSO at a flow rate of 250 nL min^{-1} is suggested. Instrument settings: Use data-dependent mode, with survey scans acquired at a resolution of 70,000 at 200 m/z and the 15 most abundant precursors selected for HCD fragmentation with an AGC target of 1×10^5 ions.
2. Label-free quantification (LFQ) analysis can be performed on the raw data using for instance MaxQuant (e.g., version 1.5.0.253) and the reference human proteome FASTA file (can be obtained from www.uniprot.org). Replicates should be collated into parameter groups to ensure matching between replicates only. Cysteine carbamidomethylation should be selected as a fixed modification and methionine oxidation as a

variable modification. Use default settings for identification and quantification; specifically, a minimum peptide length of 7, a maximum of two missed cleavage sites, and a maximum of three labeled amino acids per peptide. Through selection of the "trypsin/P" general setting, peptide bond cleavage at arginine or lysine (followed by any amino acid) is considered during in silico digest of the reference proteome. The allowed precursor and fragment ion mass tolerances are 4.5 ppm and 20 ppm, respectively. For peptide and protein identification, utilize a 0.01 false discovery rate (FDR), with "unique and razor peptides" mode selected for both identification and quantification of proteins (razor peptides are uniquely assigned to protein groups and not to individual proteins). At least 2 razor + unique peptides are suggested for valid quantification [22]. Processed data can be further analyzed using, e.g., Perseus (e.g., version 1.5.0.9) and Microsoft Excel. Peptides categorized by MaxQuant as "potential contaminants," "only identified by site," or "reverse" are filtered, and the LFQ intensities are transformed by $\log_2$. Experimental replicates are grouped, and the data filtered such that two valid LFQ values are required for the condition without any competition and at least one valid LFQ value in each of the remaining experimental groups. Proteins are further filtered to ensure that all proteins considered had been identified via a peptide spectral match, by match between runs, for the group corresponding to no competition. Missing values are imputed using default settings, and the data distribution is visually inspected to ensure that a normal distribution is maintained. Statistically significant competition is determined through the application of P2 tests, using a permutation-based FDR of 0.05 and an S0 of 2, and visualized in volcano plots.

4 Notes

1. For the treatment of live cells, the DMSO percentage should not exceed 1% to avoid increased cell death after longer incubation periods.
2. It is worth trying different concentrations, e.g., ranging from 1 to 100 μM, and to consider time-dependent labeling effects.
3. Prepare the TCEP solution freshly on the day for the click reaction. The remaining click mix stock solutions can be prepared in bulk and stored at −20 °C over 6 months.
4. Adding solvents during the methanol–chloroform precipitation results in the sample becoming cloudy due to protein precipitation. Upon vortexing, the precipitation will increase.

5. After the first centrifugation step, a delicate disk of protein precipitate will become apparent at the interphase of organic and aqueous phase. Stack a 20 μL tip on top of a 1000 μL tip and remove the supernatant gently and slowly. Then, add the first wash of methanol and gently mix by inverting without breaking the disk. After the second centrifugation step, a pellet will form at the bottom of the tube.
6. For Western blot, retain the click controls and input controls to load alongside the eluent samples to compare protein enrichment versus the original sample.
7. Monitor the vacuum carefully—liquid drops should form slowly. Turn off the vacuum when the solvent level is just above the column bed and do not let the bed run dry.
8. Pipette the sample directly onto the center of the column to avoid losing sample on the column walls.
9. Use a 1000-μL pipette tip to apply 0.6 mL eluent to the column. Use a fresh tip, insert into the column, and apply air pressure to elute the sample into an Eppendorf tube. Be careful not to aspirate any liquid into the tip.

References

1. Cook D, Brown D, Alexander R, March R, Morgan P, Satterthwaite G et al (2014) Lessons learned from the fate of AstraZeneca's drug pipeline: a five-dimensional framework. Nat Rev Drug Discov 13:419–431
2. Bunnage ME, Chekler EL, Jones LH (2013) Target validation using chemical probes. Nat Chem Biol 9:195–199
3. Stefaniak J, Huber KVM (2020) Importance of quantifying drug-target engagement in cells. ACS Med Chem Lett 11:403–406
4. Singh J, Petter RC, Baillie TA, Whitty A (2011) The resurgence of covalent drugs. Nat Rev Drug Discov 10:307–317
5. Rothweiler EM, Brennan PE, Huber KVM (2022) Covalent fragment-based ligand screening approaches for identification of novel ubiquitin proteasome system modulators. Biol Chem 403:391–402
6. Montano JL, Wang BJ, Volk RF, Warrington SE, Garda VG, Hofmann KL et al (2022) Improved electrophile design for exquisite covalent molecule selectivity. ACS Chem Biol
7. Jost C, Nitsche C, Scholz T, Roux L, Klein CD (2014) Promiscuity and selectivity in covalent enzyme inhibition: a systematic study of electrophilic fragments. J Med Chem 57:7590–7599
8. Vinogradova EV, Zhang X, Remillard D, Lazar DC, Suciu RM, Wang Y et al (2020) An activity-guided map of electrophile-cysteine interactions in primary human T cells. Cell 182:1009–26.e29
9. Robers MB, Friedman-Ohana R, Huber KVM, Kilpatrick L, Vasta JD, Berger BT et al (2020) Quantifying target occupancy of small molecules within living cells. Annu Rev Biochem 89:557–581
10. Ward JA, Pinto-Fernandez A, Cornelissen L, Bonham S, Diaz-Saez L, Riant O et al (2020) Re-evaluating the mechanism of action of alpha, beta-unsaturated carbonyl DUB Inhibitors b-AP15 and VLX1570: a paradigmatic example of unspecific protein cross-linking with Michael acceptor motif-containing drugs. J Med Chem 63:3756–3762
11. Ward JA, McLellan L, Stockley M, Gibson KR, Whitlock GA, Knights C et al (2016) Quantitative chemical proteomic profiling of ubiquitin specific proteases in intact cancer cells. ACS Chem Biol 11:3268–3272
12. Weerapana E, Speers AE, Cravatt BF (2007) Tandem orthogonal proteolysis-activity-based protein profiling (TOP-ABPP)—a general method for mapping sites of probe modification in proteomes. Nat Protoc 2:1414–1425

13. Weerapana E, Wang C, Simon GM, Richter F, Khare S, Dillon MBD et al (2010) Quantitative reactivity profiling predicts functional cysteines in proteomes. Nature 468:790–795
14. Zanon PRA, Yu F, Musacchio PZ, Lewald L, Zollo M, Krauskopf K et al (2021) Profiling the proteome-wide selectivity of diverse electrophiles. ChemRXiv
15. Hahm HS, Toroitich EK, Borne AL, Brulet JW, Libby AH, Yuan K et al (2019) Global targeting of functional tyrosines using sulfur-triazole exchange chemistry. Nat Chem Biol
16. Chung CY-S, Shin HR, Berdan CA, Ford B, Ward CC, Olzmann JA et al (2019) Covalent targeting of the vacuolar H+-ATPase activates autophagy via mTORC1 inhibition. Nat Chem Biol 15:776–785
17. Apel C, Kasper MA, Stieger CE, Hackenberger CPR, Christmann M (2019) Protein modification of lysine with 2-(2-Styrylcyclopropyl)ethanal. Org Lett 21:10043–10047
18. Ward CC, Kleinman JI, Nomura DK (2017) NHS-esters as versatile reactivity-based probes for mapping proteome-wide Ligandable hotspots. ACS Chem Biol 12:1478–1483
19. Lin S, Yang X, Jia S, Weeks AM, Hornsby M, Lee PS et al (2017) Redox-based reagents for chemoselective methionine bioconjugation. Science 355:597–602
20. Shannon DA, Gu C, McLaughlin CJ, Kaiser M, van der Hoorn RA, Weerapana E (2012) Sulfonyl fluoride analogues as activity-based probes for serine proteases. Chembiochem 13:2327–2330
21. Wang C, Abegg D, Dwyer BG, Adibekian A (2019) Discovery and evaluation of new activity-based probes for serine hydrolases. Chembiochem 20:2212–2216
22. Scheer S, Ackloo S, Medina TS, Schapira M, Li F, Ward JA et al (2019) A chemical biology toolbox to study protein methyltransferases and epigenetic signaling. Nat Commun 10:19

Chapter 15

3D Spheroid Invasion Assay for High-Throughput Screening of Small-Molecule Libraries

Kunal Karve, Stephanie Poon, Panagiotis Prinos, and Laurie Ailles

Abstract

Cancer metastasis is a complex cascade that involves the activation of cancer cell migration and invasion of the extracellular space. Cancer-associated fibroblasts (CAFs) are known inducers of cancer cell invasion. However, current in vitro invasion assays such as the Boyden chamber assay are cumbersome and low throughput. Therefore, there is an urgent need for new ex vivo, surrogate invasion assays that can faithfully recapitulate the cancer cell invasion process in vitro and are amenable to large-scale screening of small-molecule libraries in a high-throughput fashion. Here, we describe a well-established high-throughput three-dimensional (3D) spheroid invasion assay as a powerful tool to identify novel molecular targets that can potentially mediate CAF-dependent cancer cell invasion.

Key words Three-dimensional (3D), Spheroid, High-throughput, Invasion, Cancer cells, Cancer-associated fibroblasts, Small-molecule library

1 Introduction

Two-dimensional (2D) cell culture has been widely used as an in vitro model to study different aspects of physiology and disease [12, 15]. However, 2D cell culture usually fails to portray normal tissue architecture, function and disease condition [6, 20]. The recent advent of three-dimensional (3D) cell culture has gained more relevance as it more closely mimics the native environment of the tissue [17]. Adherent cells have the tendency to aggregate and when grown in specific growth conditions can form 3D multicellular spherical structures called spheroids [13]. Various methods such as hanging drop, rotating cultures, and inert scaffolds have been typically used to generate spheroids [2, 26]. Monoculture or co-culture spheroids can be derived using one or several different adherent cell types, respectively [5]. Basement membrane extract (BME) derived from the Engelbreth–Holm–Swarm (EHS) murine sarcoma line (e.g., Matrigel™) has been extensively used to

Daniel Merk and Apirat Chaikuad (eds.), *Chemogenomics: Methods and Protocols*, Methods in Molecular Biology, vol. 2706, https://doi.org/10.1007/978-1-0716-3397-7_15,

facilitate surface attachment for spheroids to grow, proliferate, and potentially invade [9]. The 3D spheroid model has tremendous potential as a powerful tool to understand and study the process of cell invasion during early development and disease conditions including cancer [14, 25].

Head and neck squamous cell carcinoma (HNSCC) is the sixth most common cancer by incidence worldwide [11, 23]. Increase in HNSCC patient morbidity and mortality is largely caused by the onset of tumor invasion and subsequent metastasis [11]. Recently, the tumor microenvironment has been shown to play a vital role in tumor metastasis of different carcinomas, including HNSCC [3, 4, 7]. Cancer-associated fibroblasts (CAFs), a cell type which resides in the tumor microenvironment, are one of the known mediators of HNSCC cancer cell invasion [8, 21]. The molecular mechanisms of how CAFs promote HNSCC cancer cell invasion remain to be fully understood. Building on the available knowledge of 3D spheroids, we have further developed, refined, and established a high-throughput 3D spheroid invasion assay as a robust tool to study CAF-dependent HNSCC cancer cell invasion [10, 18, 24]. Using a round-bottom ultra-low attachment (ULA) 384-well plate, co-cultures of Cal33 cells (an HNSCC tongue-derived cancer cell line) and HNSCC patient-derived CAFs of tongue origin generate 3D multicellular spheroids. Over a period of 3 days, in the presence of BME support and low serum conditions, CAFs promote cancer cells in the spheroids to invade the extracellular matrix unlike cancer cell monoculture spheroids. Cal33 cells stably expressing green fluorescent protein (GFP) can be employed to assess the extent of 3D spheroid invasion by using live-cell imaging techniques. Comprehensive image analysis can quantitate the magnitude of CAF-dependent Cal33 spheroid invasion. More recently, we have harnessed the power of this high-throughput assay to screen small-molecule libraries including the donated chemical probe (DCP) library from the Structural Genomics Consortium (SGC) to identify putative molecular targets, providing insights into underlying mechanisms of CAF-induced cancer cell invasion and candidate therapeutic strategies for HNSCC [16].

2 Materials

2.1 Cells

Early passage Cal33 cells or CAFs tested mycoplasma negative and were approximately 70% confluent at the time of harvest for the 3D spheroid invasion assay setup. CAFs were derived from primary HNSCC samples as described in Principe et al [19].

1. Fluorescent protein stably expressing cancer cells (in our case GFP expressing Cal33 cells/Cal33-GFP) (*see* **Notes 1** and **2**).
2. HNSCC patient-derived cancer-associated fibroblasts (*see* **Notes 3** and **4**).

2.2 Cell Culture Reagents

1. 1X phosphate-buffered saline/PBS: Gibco, Cat: 14190-144.
2. Iscove's modified Dulbecco media/IMDM: Wisent Inc., Cat: 319-105-CL.
3. Fetal bovine serum (FBS): Gibco, Cat: 12483-020.
4. Complete cell culture growth media: IMDM with 10% FBS.
5. Cell culture invasion assay media: IMDM with 2% FBS (*see* **Note 5**).
6. 0.25% trypsin: Gibco, Cat: 25200072.
7. Growth factor-reduced (GFR)-BME such as Cultrex BME: Biotechne, Cat: 3533-010-02P or Matrigel™: Corning, Cat: 354230.
8. 0.4% trypan blue reagent: Gibco, Cat: 15250061.
9. DMSO and donated chemical probes (DCP) library in a 384-well Echo liquid handler compatible master plate: Labcyte, Cat: LP-0200, obtained from the Structural Genomics Consortium (SGC): https://www.thesgc.org/donated-chemical-probe.

2.3 Cell Culture Accessories

1. Cell culture biosafety cabinet and incubators (*see* **Note 6**).
2. Microscope and hemacytometer: Millipore Sigma, Cat: Z359629.
3. Temperature-controlled plate centrifuge.
4. Echo 650 T liquid handler: Beckman Coulter.
5. Sterile 384-well U-bottom ultra-low attachment (ULA) plate with lid: Sbio, Cat: MS-9384UZ.
6. Sterile 384-well daughter plate with lid: Nunc, Cat: 242757.
7. 16-channel pipettes, i.e., 384-well pipettes and sterile pipette tips.
8. Cell culture sterile reagent reservoirs.
9. Ice buckets and centrifuge-balanced plates.
10. Kimwipes: Kimberly-Clark Professional™, Cat: 06-666A.

2.4 Imaging and Analysis

1. S3 live-cell analysis system: Sartorius.
2. Incucyte spheroid module: Sartorius.
3. ILASTIK software for image analysis.
4. ImageJ software for image analysis.
5. Data storage system.
6. Microsoft Excel and PowerPoint.
7. Graph-Pad Prism.

3 Methods (Fig. 1)

3.1 Pre-setup Day: − 2

Verify that the Cal33-GFP and CAFs are in a healthy growing phase. Change media with fresh complete growth media.

3.2 Setup Day: − 1

3.2.1 Preparation of Cells

1. Check that the Cal33-GFP cells and CAFs are approximately 65–80% confluent.
2. Aspirate the complete growth media: IMDM with 10% FBS and wash the cells with 1X PBS.
3. Dissociate cells with 0.25% trypsin and quench the trypsin with complete growth media.
4. Collect the dissociated cell slurry in 15-ml tubes, centrifuge at 250 *g* for 5 min at room temperature.
5. Discard the spent media.
6. Resuspend the cell pellets with the appropriate amount of cell culture invasion assay media: IMDM with 2% FBS.
7. Take a small aliquot of cells and mix in 1:1 volume ratio with trypan blue.

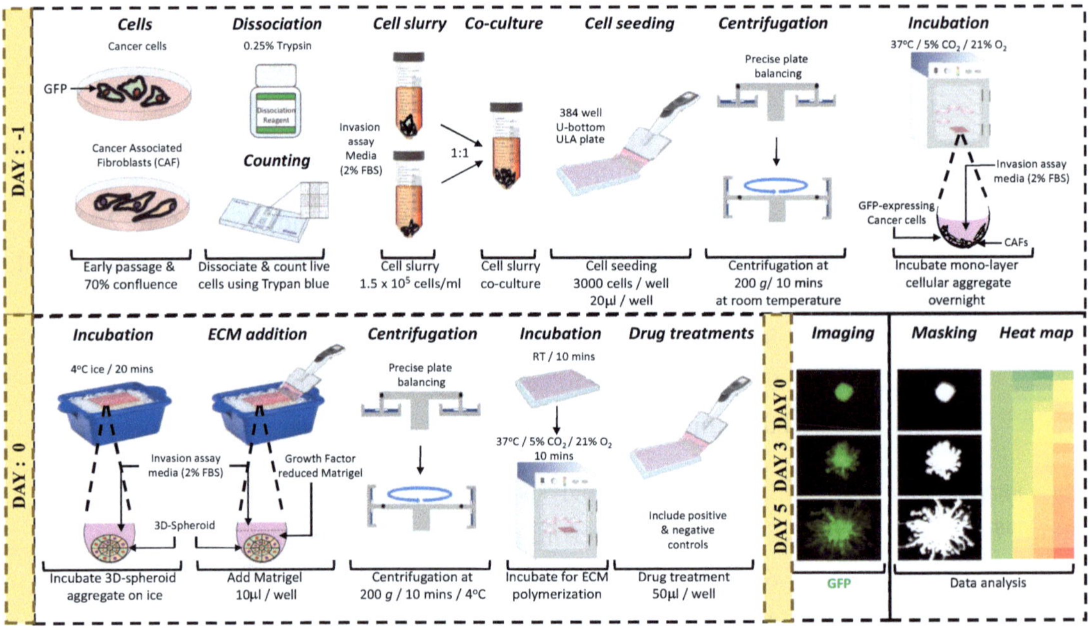

Fig. 1 3D spheroid invasion assay protocol: Diagrammatic representation of the key steps involved in the setup of 3D spheroid invasion assay. Top left to right (day: −1) involves harvesting Cal33-GFP (cancer cells) and CAFs and through subsequent steps seeding them in a 384-well U-bottom ultra-low attachment plate. Bottom left (day: 0) requires addition of GFR-BME as an ECM matrix followed by drug treatment. Bottom right (days 0, 3, and 5) employs imaging, image analysis techniques, data analysis, and data plotting

8. Count the live cells using a hemacytometer. Make sure to use the dilution factor 2 in the cell count calculation to compensate for the dilution in trypan blue.
9. Prepare separate cell slurries of Cal33-GFP cells and CAFs with a cell density of 1.5×10^5 cells per ml using cell culture invasion assay media.

3.2.2 Seeding of Cells

1. Along with individual Cal33-GFP and CAF slurries, prepare the co-culture cell slurry mix in a separate tube using an aliquot of the individual slurries in a 1:1 Cal33-GFP to CAF ratio (*see* **Note** 7).
2. Using a 16-channel pipette, transfer 20 μl per well (3000 cells per well) cell slurry as per requirement (*see* **Notes 8 and 9**). Plan to have a minimum of 3 replicate wells (we typically use 6) per condition to be tested.
3. Along with transferring 20 μl per well of the co-culture slurries, make sure to include wells of Cal33-GFP and CAF monoculture slurries (20 μl per well) that would act as negative controls (*see* **Note 10)**.
4. Fill the empty wells with 80 μl of 1X PBS.
5. Using a balance plate, centrifuge the plate seeded with cells at 200 *g* at room temperature for 3–5 min (*see* **Note 11**).
6. Check that the cells have settled at the bottom center of the U-shaped well after centrifugation.
7. Incubate the plate in a 37 °C normoxic (21% O_2, 5% CO_2) cell culture incubator overnight (*see* **Notes 6, 12,** and **13**).

3.2.3 Preparation for the Next Day

1. Calculate the required amount of GFR-BME required (10 μL per well seeded) for the 3D spheroid invasion assay and thaw it overnight in a 4 °C refrigerator (*see* **Note 14**).

3.3 Setup Day: 0

3.3.1 Addition of the Cell Matrix

1. Check the wells for the formation of 3D spheroid aggregates under the microscope (*see* **Notes 15** and **16**).
2. Keep the plate with the spheroids on ice for 15–20 min.
3. Keep the cell culture reagent reservoir submerged in ice up to brim to cool down.
4. After 15–20 min add GFR-BME to the reservoir.
5. Using a 16-channel pipette, gently transfer into the media 10 μl of cold GFR-BME per well of the 384-well plate kept on ice to get a final BME concentration of approximately 33% (*see* **Notes 17** and **18**).
6. Centrifuge the plate at 200 *g* for 10 min at 4 °C. Rotate the plate by 180° and centrifuge for another 10 min (*see* **Notes 11** and **19**).

7. Observe the 3D spheroid under the microscope and then keep the plate at room temperature for 10–15 min.
8. Keep the plate in the normoxic incubator at 37 °C for 10–15 min for the BME to polymerize (*see* **Notes 6, 12,** and **13**).

3.3.2 Drug Treatments

The donated chemical probe (DCP) library consists of potent, selective, and broadly characterized small-molecule chemical modulators of protein function donated by 8 global pharmaceutical companies to promote open science and explore novel biology. This well-characterized chemical probe set is freely available from the Structural Genomics Consortium (SGC), a public–private partnership for the discovery of new medicines. More details can be found here: https://www.sgc-ffm.uni-frankfurt.de/chemProbes. The DCP library chemical probes were obtained in a 384-well Echo compatible master plate at a stock concentration of 10 mM (*see* **Notes 20** and **21**).

1. Aliquot the stock DCP probes from the 384-well master plate into multiple 384-well daughter plates using an Echo acoustic liquid handling system.
2. Echo liquid handlers use acoustics to accurately transfer small volume of liquids as low as 2.5 nl from the 384-well Echo master plate to multiple replicates of 384-well daughter plates in a high-throughput fashion.
3. To perform Echo-mediated chemical probe transfer, accurately curate a detailed "transfer protocol."
4. The "transfer protocol" should consist information such as source plate name, destination/daughter plate name, drug name, row and column number, source well, destination well, and transfer volume.
5. The "transfer protocol" step for Echo helps us to design the daughter plate layout, for the controls (DMSO) and chemical probes with multiple technical replicates which are well-distributed across the daughter plate (*see* **Notes 22** and **23**).
6. Use assistance from an expert who is trained to operate the Echo liquid handler to perform the chemical probe transfers.
7. Using the Echo, dispense tiny droplets of 8.5 nl volume containing each of the chemical probes into the individual wells of the 384-well daughter plate.
8. The 8.5 nl volume of each chemical probe was calculated keeping in mind that there will be cell seeding media and BME in the spheroid assay U-bottom well prior to treatment which can further dilute the probe. Extra volume was also included to eliminate error due to pipetting or during transfer from the daughter plate to 3D spheroid assay ULA U-bottom plate at the time of the treatment.

9. Prepare the daughter plates with the chemical probes beforehand, store them at -80 °C, and use when needed for cell treatments.
10. Add 55 μl of cell culture invasion assay media: IMDM with 2% FBS, to each well of the daughter plate, to make a 1:10,000 dilution of the stock concentration for a final working concentration of 1 μM.
11. Match the layout of the chemical probes and controls in the wells of the 384-well daughter plates to the treatment pattern of the 384-well 3D spheroid assay ULA U-bottom plate.
12. Transfer 50 μl of the invasion assay media containing 1 μM of each probe and the DMSO control from the daughter plate directly to the 3D spheroid ULA U-bottom assay plate using a 16-channel pipette, after polymerization of the GFR-BME (part −3.3.1).
13. Maintain sterile conditions throughout the process.
14. Keep the plate in a 37 °C normoxic cell culture incubator (*see* **Notes 6, 12, and 13**).

3.4 Imaging Days: 0, 3, and 5 and Analysis

3.4.1 Spheroid Imaging

1. Image the spheroids using an imaging system that captures both GFP fluorescence and brightfield images (*see* **Note 24**).
2. The S3 live-cell analysis system: Sartorius, with the spheroid module installed captures good images in two dimensions.
3. Acquire the spheroid images on days 0, 3, and 5, respectively (*see* **Note 25**).

3.4.2 Image Analysis

Image Masking Using ILASTIK

1. Download the most recent version of the ILASTIK software (www.ilastik.org) [1].
2. Open ILASTIK software.
3. Select "Pixel Classification module."
4. Save the project; i.e., give it a project name.
5. Use GFP florescence images for cell masking purposes.
6. In the input data (first tab on the left), drag 5–10 images that represent your image set to the window on the upper right.
7. Go to the second tab—"feature selection"—click the "select features" button and select all the options (all are checked and turn green). Press "OK."
8. Click the third tab—"training." Relabel "label 1" as "spheroid" and "label 2" as "background."
9. Using the paintbrush tool, draw lines on the spheroid to designate what are spheroid (with yellow brush) and background (with blue brush). Check (tick) "probability" and "segmentation" tabs and do "live update" to see spheroid masking. If

required, change the lines (yellow and blue) designating spheroid and background using an eraser tool to make masking better/more accurate.

10. Go to the fourth tab, "Prediction export." Select "choose export image settings" option. A dialog box opens, check "Transport to axis order" and write "cyx instead of yxc." In the output file info, select the format as "tiff sequence." For the "file" option, select the directory where you would like to store the binary masked images. Press "ok."
11. Choose the fifth tab, "batch processing." Drop off all the images for which you want to get the binary images. Press "process all files."
12. The software will take 10–15 min to generate the binary (black and white) masked images. Once the processing is done, close the ILASTIK software.
13. Locate the folder where the binary images are saved.
14. The folder will contain two mask probabilities per image. One with the white-colored spheroid and black background and the other with the colors reversed.
15. Keep all mask predictions where the spheroid is in white color and the background is in black and delete the other predictions.
16. Make sure to visually compare all the mask predictions to the respective source GFP fluorescence images to ensure that the image masking done by batch processing method is precise and accurate.

Cell Segmentation and Image Analysis Using ImageJ FIJI

1. Download the most recent version of the ImageJ FIJI software (imagej.net/software/fiji/) and open [22].
2. Open one masked image of the entire set in the FIJI software.
3. First, to set image scale, click the "Analyze" dropdown menu and select "set scale".
4. A dialog box will open. If the source image was generated using the Incucyte imaging system at 10X magnification, use the following parameter settings to set scale of the image. For "Distance in pixels," put "1"; for "Known distance," put "3.05." Adjacent to "Pixel aspect ratio," put "1," and for "Unit of length," put "μm." Select the "Global" option check box and click "OK." If the image is from a different source, then one may need to modify the image scale parameters accordingly.
5. Close the opened masked image which was used to set image scale.
6. Go to the "Analyze" dropdown menu and click "Set Measurements" and select "Area," "Shape descriptions," and "Display label" in the check boxes. Click "OK."

7. Click the "Process" dropdown menu and through the "Batch" option select "Macro."
8. In the "Input" option, select the file directory where the masked images are stored. Keep the "Output" option empty.
9. In the empty space, paste the following macro code for batch processing of the images (*see* **Notes 26 and 27**):

```
setAutoThreshold("Default dark");
//run("Threshold...");
run("Analyze Particles...", "size=80000.00-Infinity show=O-
    verlay display add in_situ");
```

10. Click "Test" to run the code on one random image from the input file to confirm it works. Click "Process" button to proceed with all images in the Input folder.
11. Values for spheroid area and circularity will be generated in a new window.
12. These values can be copied into Microsoft Excel for further analysis.

3.4.3 Data Visualization and Statistics

1. Arrange representative 3D spheroid GFP fluorescence and phase contrast images for different treatments and control groups in a panel in a Microsoft PowerPoint file.
2. Copy the values obtained for spheroid area and inverse circularity into a spreadsheet, reorganize them into groups based on cell type and treatment conditions, analyze, and then plot into graphs or heat maps (*see* **Note 28**).
3. Plot the data as a fold change in the area or inverse circularity at day 3 or 5 relative to day 0 control.
4. Perform statistic tests such as Student's t-test or ANOVA to estimate the significance of change across different treatment conditions.

4 Notes

1. Cal33-GFP cells in 2D culture were grown in complete cell culture growth media: IMDM with 10% FBS, at 37 °C in a normoxic cell culture incubator.
2. Check the GFP expression in the Cal33 cells under a fluorescence microscope prior to 3D spheroid invasion assay setup. Poor fluorescent protein expression in cells may lead to difficulties during the image analysis steps; this can prevent ILASTIK from separating invasive arms of the spheroid from the background.

3. Our HNSCC patient-derived CAF lines in 2D culture were grown in complete cell culture growth media: IMDM with 10% FBS, at 37 °C in a hypoxic (2% O_2 and 5% CO_2) cell culture incubator. CAFs can also be grown in a normoxic incubator, but growth rate will be slightly slower.
4. We have tested multiple HNSCC patient-derived CAF lines in the 3D spheroid invasion co-culture assays and observed that all CAF lines tested promote Cal33-GFP 3D spheroid invasion, but at varying degrees of invasion and/or invasion patterns. During initial assay standardization, it is advised to test multiple available HNSCC CAF lines. Post-standardization, for the high-throughput 3D spheroid invasion assays pick the CAF line that produces the most pronounced phenotypic change in terms of 3D spheroid invasion in Cal33-GFP and CAF co-cultures.
5. Make fresh cell culture invasion assay media: IMDM with 2% FBS, on the day of cell seeding.
6. Maintain sterile conditions to minimize chances of contamination. The full 3D invasion assay setup is performed in the biosafety cabinet.
7. The CAF to cancer cell ratio in the co-cultures can be increased from 1:1 (used in our case) to 2:1 or 5:1 for other cancer cell types and CAF lines to achieve the maximum cancer cell 3D spheroid invasion phenotype. Note that a 5:1 co-culture ratio may skew the circularity measurements of day 0 images because of the sparse distribution of fluorescently labeled cancer cells within the spheroid.
8. Some cell types may form larger spheroids due to large cell size which may become an issue for imaging as the spheroid may be larger or invade outside the field of view. For Cal33, we used 3000 cells per well of a 384-well plate, but for other cell types, one may need to seed a lower number of cells to generate spheroids.
9. Avoid using wells in the periphery of the plate. Balance them with equivalent volume of 1X PBS during different steps of the assay setup.
10. In this particular protocol, Cal33-GFP and CAF monoculture-derived 3D spheroids act as a negative spheroid invasion assay control as they do not invade the matrix unlike the co-culture-derived 3D spheroids. However, other cancer cell lines and even some CAF lines may demonstrate invasion as a monoculture spheroid.
11. Ensure there is a proper balance plate for each centrifugation step. Perform fine balancing by maintaining similar volumes using 1X PBS in the balance plate.

12. To avoid media evaporation during incubation at 37 °C, keep the 384-well plate in a secondary container with sterile distilled water and ensure the water pan of the tissue culture incubator is not empty. Also, ensure the water pan of the incubator that holds the Incucyte S3 imaging system is not empty.
13. For the entire duration of the 3D spheroid invasion assay wherever necessary, the U-bottom invasion assay plates were incubated in a 37 °C normoxic cell culture incubator.
14. Keep the BME-containing vial buried in ice in a bucket that is kept inside a 4 °C refrigerator overnight for a complete and an even thaw.
15. It is important to test for spheroid generation of cancer cell lines prior to starting, as not all will self-aggregate. A good way to confirm whether an aggregate has formed is to pipette the media containing the spheroid up and down with a tip larger than the size of the spheroid and afterward looking under the microscope to evaluate whether the aggregate has kept its shape.
16. Spheroid formation for some cell types may take longer than a day. In this case, incubate longer and adjust time of addition of the GFR-BME accordingly.
17. Avoid air bubbles while adding the BME to each well.
18. Gently add BME against the walls distributing it on both sides of the well so the spheroid remains centered in the well.
19. After BME addition to each well, if centrifuging twice does not make the spheroid land in the center of the well, then centrifuge a few more times.
20. Prior to performing high-throughput chemical probe screens using the 3D spheroid invasion assay, it is important to test if the chemical probes have cytostatic or cytotoxic effects on the cancer cells or CAF lines at the working concentration or dose (1 μM in our case) by performing 2D cell viability assays. The goal of the 3D spheroid invasion assay is to identify chemical probes that curb invasion of the cancer cells. Any cytotoxic effects may lead to false positive hits.
21. During 2D cell viability assays and 3D spheroid invasion assays, use the appropriate chemical probe structural analog negative control to eliminate the possibility of off-target effects of a chemical probe.
22. Determine the DMSO toxicity to the cells beforehand to avoid false positives. We kept DMSO at a concentration of 0.01% during chemical probe treatments, which did not affect the viability of Cal33 cells or CAFs.

23. Use appropriate assay-positive controls.
24. Gently wipe the bottom of the plate with a kimwipe prior to imaging to avoid fuzzy images.
25. The day of imaging can be adjusted to represent the best phenotypic change in terms of 3D spheroid invasion.
26. In the macro code for the image segmentation step, the threshold is set to "Default dark." This segmentation method may not always work optimally for all images. This may be due to various factors, such as variation in different experiments, the extent or pattern of invasion, and sharpness of the source image. There are multiple segmentation methods that can be applied to the image analysis in the ImageJ FIJI software. To select which segmentation method is appropriate, a segmentation test on one or more binary images should first be performed. For this, open one binary image in ImageJ FIJI. Set the image "scale" and "measurements" as described earlier. Click the "Image" tab on the top of the ImageJ FIJI software. In the dropdown menu, select "Adjust" and click on the "Threshold..." option. A dialog box opens. On the right-hand side dropdown menu in the dialog box, select "red." On the left-hand side dropdown menu, multiple segmentation options will appear (e.g., MaxEntropy, RenyiEntropy). Select the segmentation method that most accurately detects the spheroid (this is seen as a red mask over the image). Check boxes for "Dark background" and "Don't reset range." To check if the area and the circularity measurement output for the selected image are appropriate to the segmentation, click the "Analyze" dropdown menu and select, "Analyze particles...". A dialog box will open. For "Size (μM^2)," put "0-infinity," for "circularity," put "0.00–1.00," and for "Show," select "nothing." Select 'Display results," check box, and click "OK." A new results window will open with the corresponding measurements for Area and Circularity. This can be tested for multiple sample images to designate a segmentation method that later can be applied for batch processing using the macro code. For example, if "RenyiEntropy" is appropriate for image segmentation, replace "Default dark" in the macro code with "RenyiEntropy dark."
27. In the macro code, the objects to be analyzed in ImageJ FIJI software are determined by the set threshold for the spheroid size. Any object in the binary image above the threshold is considered a spheroid. In our case, the threshold for spheroid size is "Size = 80000.00-infinity" which means any object below size 80000 is ignored. The threshold value for spheroid size in the macro code can be increased or decreased based on

the spheroid size. The spheroid size may vary somewhat across experiments or for different cell types. For example, for smaller spheroids the threshold can be reduced to “size = 60000.00” or below.

28. The values obtained for spheroid circularity using ImageJ FIJI software are represented as inverse circularity (i.e., invasiveness) to depict an increase in invasiveness of the co-culture spheroids over Cal33-GFP-derived monoculture spheroids. Inverse circularity is simply 1/circularity output value.

Acknowledgements

Thanks to Dr. Christina Karamboulas, Dr. Ngoc Bui and Jalna Meens for their help and training in the Ailles laboratory. Thanks to Dr. Claudia Tredup and Dr. Bennedict Tillman-Berger from SGC-Frankfurt and Dr. Suzanne Ackloo and Albina Bolotokova from SGC-Toronto laboratories for chemical probe management, sourcing, and distribution. Tissues used to generate the CAF lines were collected from HNSCC patients who underwent surgical resection of their tumors at the University Health Network (UHN) and provided written consent under UHN Research Ethics Board approval (REB no. 12-5639). Thanks to Drs. Rodger Tiedemann, Shane Harding and Robert Rottapel for generously allowing us to use their Incucyte imaging system. Thanks to Dr. Carman K M Ip and Dr. Troy Ketela at the Princess Margaret Genomics Centre (PMGC) for helping to make the chemical probe daughter plates using the Echo liquid handler. The authors appreciate the excellent research environment and the research facilities provided by the Princess Margaret Cancer Centre and the SGC. Kunal Karve was supported by a Mitacs Accelerate fellowship. Stephanie Poon was supported by Ontario Student Opportunity Trust Funds endowed by the Princess Margaret Cancer Center, Ontario Institute for Cancer Research, the University of Toronto and the Province of Ontario. This work was partly funded by a research contract awarded to Drs. Ailles and the SGC from Merck KGaA, Darmstadt, Germany, and by an Investigator Award from the Ontario Institute for Cancer Research. The SGC is a charity (register number 1097737) that received funding from AbbVie, Bayer AG, Boehringer Ingelheim, Bristol Myers Squibb, Genentech, Genome Canada through Ontario Genomics Institute [OGI-196], Janssen, Merck KGaA (aka EMD in Canada and USA), Pfizer, Takeda, and the Innovative Medicines Initiative 2 Joint Undertaking (IMI2 JU; EUbOPEN grant 875510) from the European Union and European Federation of Pharmaceutical Industries and Associations (EFPIA).

References

1. Berg S, Kutra D, Kroeger T et al (2019) Ilastik: interactive machine learning for (bio)image analysis. Nat Methods 16:1226–1232
2. Cacciamali A, Villa R, Dotti S (2022) 3D cell cultures: evolution of an ancient tool for new applications. Front Physiol 13:836480
3. Chaffer CL, Weinberg RA (2011) A perspective on cancer cell metastasis. Science 331: 1559–1564
4. Elmusrati A, Wang J, Wang CY (2021) Tumor microenvironment and immune evasion in head and neck squamous cell carcinoma. Int J Oral Sci 13:24
5. Fennema E, Rivron N, Rouwkema J et al (2013) Spheroid culture as a tool for creating 3D complex tissues. Trends Biotechnol 31: 108–115
6. Fontoura JC, Viezzer C, Dos Santos FG et al (2020) Comparison of 2D and 3D cell culture models for cell growth, gene expression and drug resistance. Mater Sci Eng C Mater Biol Appl 107:110264
7. Friedl P, Alexander S (2011) Cancer invasion and the microenvironment: plasticity and reciprocity. Cell 147:992–1009
8. Hanahan D (2022) Hallmarks of cancer: new dimensions. Cancer Discov 12:31–46
9. Hughes CS, Postovit LM, Lajoie GA (2010) Matrigel: a complex protein mixture required for optimal growth of cell culture. Proteomics 10:1886–1890
10. Hussain A, Voisin V, Poon S et al (2020) Distinct fibroblast functional states drive clinical outcomes in ovarian cancer and are regulated by TCF21. J Exp Med 217
11. Johnson DE, Burtness B, Leemans CR et al (2020) Head and neck squamous cell carcinoma. Nat Rev Dis Primers 6:92
12. Kapalczynska M, Kolenda T, Przybyla W et al (2018) 2D and 3D cell cultures – a comparison of different types of cancer cell cultures. Arch Med Sci 14:910–919
13. Laschke MW, Menger MD (2017) Life is 3D: boosting spheroid function for tissue engineering. Trends Biotechnol 35:133–144
14. Li S, Zhang Z, Han L (2020) 3D spheroids propel tumor characterization. Trends Cancer 6:622–624
15. Martinotti S, Ranzato E (2020) Scratch wound healing assay. Methods Mol Biol 2109:225–229
16. Muller S, Ackloo S, Arrowsmith CH et al (2018) Donated chemical probes for open science. elife 7
17. Pampaloni F, Reynaud EG, Stelzer EH (2007) The third dimension bridges the gap between cell culture and live tissue. Nat Rev Mol Cell Biol 8:839–845
18. Poon S, Ailles LE (2022) Modeling the Role of cancer-associated fibroblasts in tumor cell invasion. Cancers (Basel) 14
19. Principe S, Mejia-Guerrero S, Ignatchenko V et al (2018) Proteomic analysis of cancer-associated fibroblasts reveals a paracrine role for MFAP5 in human oral tongue squamous cell carcinoma. J Proteome Res 17:2045–2059
20. Rawal P, Tripathi DM, Nain V et al (2022) VEGF-mediated tumour growth and EMT in 2D and 3D cell culture models of hepatocellular carcinoma. Oncol Lett 24:315
21. Sahai E, Astsaturov I, Cukierman E et al (2020) A framework for advancing our understanding of cancer-associated fibroblasts. Nat Rev Cancer 20:174–186
22. Schneider CA, Rasband WS, Eliceiri KW (2012) NIH image to ImageJ: 25 years of image analysis. Nat Methods 9:671–675
23. Vigneswaran N, Williams MD (2014) Epidemiologic trends in head and neck cancer and aids in diagnosis. Oral Maxillofac Surg Clin North Am 26:123–141
24. Vinci M, Box C, Eccles SA (2015) Three-dimensional (3D) tumor spheroid invasion assay. J Vis Exp:e52686
25. Wang Y, Jeon H (2022) 3D cell cultures toward quantitative high-throughput drug screening. Trends Pharmacol Sci 43:569–581
26. Yamada KM, Cukierman E (2007) Modeling tissue morphogenesis and cancer in 3D. Cell 130:601–610

Chapter 16

Live-Cell High-Throughput Screen for Monitoring Autophagy Flux

Sara Cano-Franco, Hung Ho-Xuan, Lorene Brunello, and Alexandra Stolz

Abstract

Autophagy is a cellular process implicated in the renewal of cellular components and the maintenance of cellular hemostasis and therefore associated with various types of diseases. In addition, autophagy belongs to the stress response pathways and is frequently activated by chemical compounds harboring characteristics of cell toxicity. High-throughput screens analyzing autophagy flux are therefore applied in both, the field of compound identification for targeting autophagy and compound characterization for analyzing compound toxicity. In this chapter, we describe a live-cell, fluorescent-based, high-throughput screening method in 384-well format for the fast and accurate measurement of autophagy flux over time suitable for academic research, pharmacological applications, and drug discovery.

Key words Autophagy, GFP-LC3-RFP-LC3ΔG, Fluorescence, Incucyte

1 Introduction

Awarded with the Nobel Prize in 2016, autophagy is the center of a continuously growing field of research and drug development. It is a quality control and stress response pathway with an impact on the development of several human diseases [1–3]. The autophagy pathway is also highly interconnected with other cellular quality control and stress response pathways [4]. While a lot of central players and regulators of autophagy are meanwhile known and well understood, many fine-tuning factors are yet to be discovered. This makes a prediction of the impact of certain conditions and chemical treatments on autophagy flux extremely difficult. Therefore, cellular assays for an unbiased quantitative analysis of autophagy flux are becoming a standard requirement in a number of research fields.

Authors: "Sara Cano-Franco" and "Hung Ho-Xuan" have equally contributed to this chapter.

Supplementary Information The online version contains supplementary material available at https://doi.org/10.1007/978-1-0716-3397-7_16.

Daniel Merk and Apirat Chaikuad (eds.), *Chemogenomics: Methods and Protocols*, Methods in Molecular Biology, vol. 2706, https://doi.org/10.1007/978-1-0716-3397-7_16,

Each assay for analyzing autophagy developed over the years has its own pros and cons [5]. However, the development of protein-based fluorescent autophagy reporters for general and selective autophagy pathways has certainly opened the door for high-throughput screening (HTS) approaches: protein levels of central players and their degradation in the lysosome via autophagy are quantified and monitored with the help of fluorescent protein tags. The low pH in the lysosome plays an important role in this setting, as it allows the efficient use of pH-sensitive fluorophores in ratiometric assays comparing cytosolic and lysosomal localization, respectively [5, 6]. The most commonly used reporter protein for measuring general autophagy flux is LC3B, a member of the Atg8 protein family [7–9]. Reporters for selective autophagy pathways (e.g., ER-phagy, mitophagy, and lysophagy) are either based on selective autophagy receptors of the pathway or proteins localized to specific organelles or structures: SQSTM1/p62 [10]; mCherry-eGFP-FAM134B [11], mCherry-eGFP-RAMP4 [12, 13], mKeima-RAMP4 [14], ssRFP-GFP-KDEL [15, 16], mCherry-GFP-REEP5 [17], mt-mKeima [18], mito-SRAI [19], mCherry-GFP-FIS1 [20, 21], RFP-EGFP-LGALS3, or mKeima-LGALS3 [22].

In this chapter, we describe in detail a 384-well format autophagy flux assay using a stable reporter cell line expressing the GFP-LC3-RFP-LC3ΔG reporter from the Mizushima Lab [8]. In addition, we provide an analysis file for basic quality control and semi-automated analysis of the data.

2 Materials

This chapter refers to 384-well plate format. Laboratories without/limited access to automated liquid handlers and dispensers can perform the screen with reduced throughput in 96-well plates using multichannel pipettes.

2.1 Reporter Cell Line

1. Plasmid pMRX-IP-GFP-LC3-RFP-LC3ΔG (Addgene plasmids #84572 [8]; *see* **Note 1**).
2. Cell line of your choice (*see* **Note 2**).

Here, RPE-1 cells stably expressing GFP-LC3-RFP-LC3ΔG (single clone; FACS sorted for a uniform medium expression) are used [23].

2.2 Preparation of Compound Plates

1. Echo® 550. *Alternative*: (multichannel) pipette.
2. V-shaped 96-well plates (low compound binding).
3. Echo® source plates (LP-0200, Labcyte®).

4. Control compounds: DMSO, 250 nM Torin 1 (stock: 1 mM), 321 nM bafilomycin A1 (stock: 321 μM) (*see* **Note 3, 4**).
5. Chemical library of interest (*see* **Note 5**).

2.3 Cell Seeding and Screen

1. Standard material and equipment for culturing cell lines.
2. Automated cell counter.
3. Thermo Scientific™ Multidrop™ Combi dispenser. *Alternative*: multichannel pipette (8 or 12 channels; range 10–100 μl).
4. F-shaped 384 well plates with clear bottom (*see* **Note 6**). *Alternative*: 96-well plate with clear bottom.
5. Liquidator™ 96 Manual Pipetting (range 5–200 μl). *Alternative*: multichannel pipet (8 or 12 channels; range 10–100 μl).
6. Incucyte® (Sartorius). *Alternative*: fluorescent plate reader able to detect GFP and RFP.

3 Methods

Use sterile material and work as sterile as possible. Whenever possible, carry out procedures under a sterile bench. Use pre-warmed media and solutions to avoid temperature stress for cells.

3.1 Cell Seeding (Day 1)

1. Seed 50 μl of cells (17.500 cells/ml) per well in the 384-well screening plate with the help of the Multidrop™ (*see* **Notes 7–10**).
2. Leave the plate for 10–15 min under the hood at RT (*see* **Note 11**).
3. Grow cells for 20–24 h to 20% confluency (*see* **Note 12**).

3.2 Preparation of Compound Plates (Days 1 and 2)

This protocol was optimized to make screens most robust and data comparable within the usual academic setting (different users, medium-to-low automatization). Therefore, we implemented the preparation of compounds as an individual step, despite the need for extra material (96-well plates, tips). Facilities with high automatization and/or highly trained staff may avoid this extra step (*see* **Note 13**). It is recommended to prepare the compound plates on day 1 without media and store them at −20 °C till day 2 to avoid possible delays in the treatment procedure due to technical issues that may occur while using the Echo® 550.

1. Use the Echo® 550 to prepare one V-shaped 96-well plate with screening compounds (2x final screening concentration; 170 μl volume) according to the plate map (Fig. 1; *see* **Note 14**).

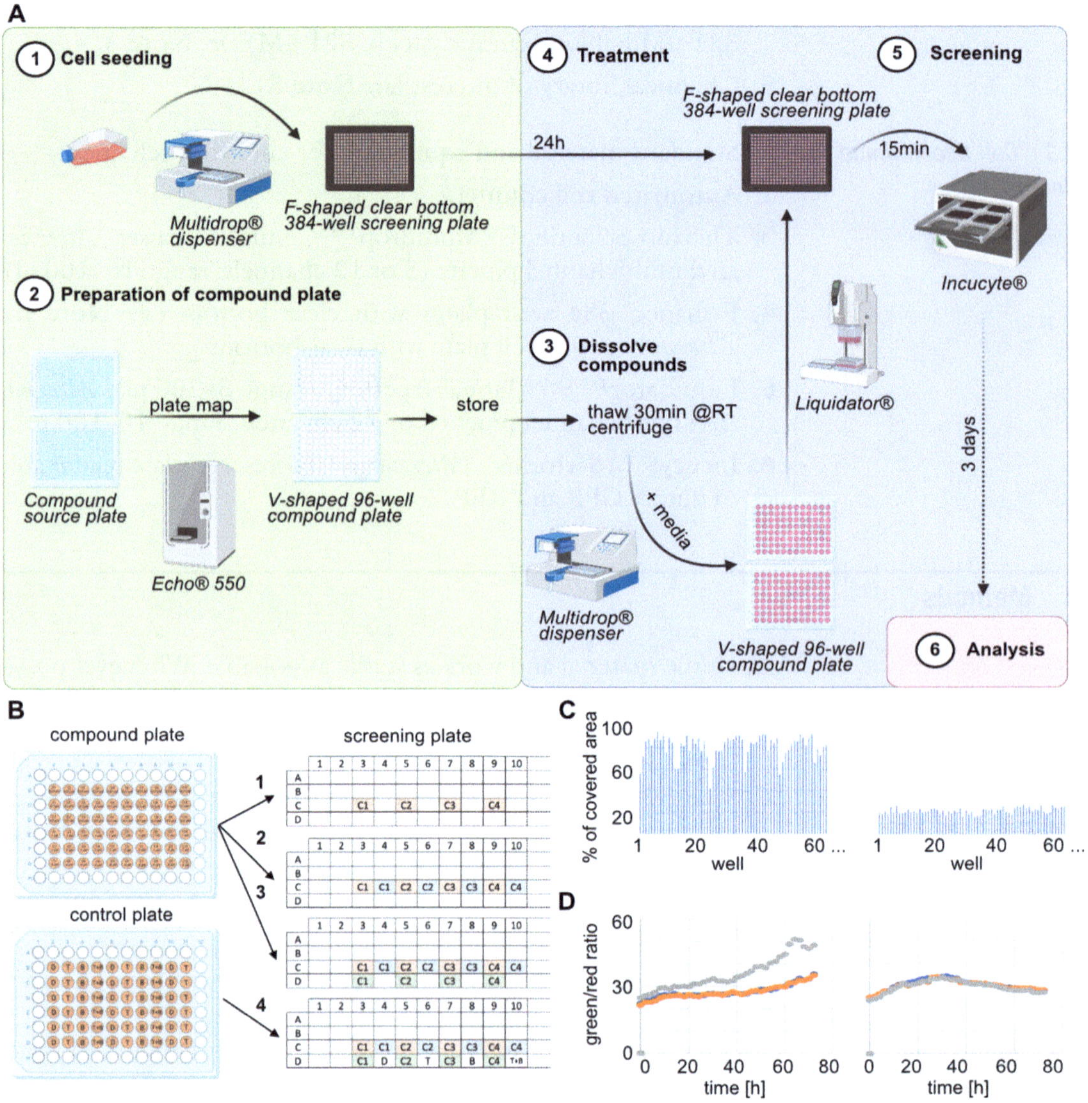

Fig. 1 Screening procedure and basic quality control. (**a**) Schematic representation of the workflow. (**b**) Compounds are transferred manually in four steps from compound plates (96-well) to the screening plate (384-well) to create islands of triplicates and one control compound. (**c**) Initial equal seeding is essential for reproducibility and comparability. Left panel: unequal and too high seeding density; right panel: equal seeding at 20% (**d**) Representative graphs of replicates passing (right panel) and not passing (left panel) basic quality control. In case of a clear outliner, individual replicates should be manually curated to avoid false-positive hits

2. Use the Echo® 550 to prepare one V-shaped 96-well plate with control compounds (2x final screening concentration; 170 μl (70 μl) volume; lasts for 3 (1) screening plates) according to the plate map (Fig. 1; *see* **Note 4, 15**).
3. Store both compound plates at −20 °C.

3.3 Screening (Day 2)

1. On day 2, thaw prepared compound plates for 30 min at RT and then centrifuge for 3 min at 1500 rpm. Add 170 μl of medium to each well (including the outer wells without compound) with the Multidrop™. Incubate plates at 37 °C for >30 min.
2. Prepare the screening program on the Incucyte® software (*see* **Note 16**).
3. Transfer 3x 50 μl of screening compound into the respective wells of the 384-well screening plate using the Liquidator™ 96 (Fig. 1a; *see* **Note 10**).
4. Transfer 50 μl of control compound into the respective well of the 384-well screening plate using the Liquidator™ 96 (Fig. 1a).
5. Check for potential bubbles and transfer the screening plate into the Incucyte® (*see* **Note 17**).
6. Wait for 15 min and then start with the first screening time point (*see* **Note 18**).

3.4 Data Analysis and Basic Quality Control

1. Set up a basic analysis for confluency (phase; percent of covered area) and the fluorescence signal of the reporter (green, red; total integrated intensity, GCU x μm2/image or RCU x μm2/image, respectively) following the instructions of the Incucyte® software (*see* **Note 19**).
2. Extract data as an Excel file and transfer values into the respective "raw data" taps of the provided analysis sheet (*see* **Note 20**).
3. The ratio of green and red fluorescence may be used to compare autophagy flux between individual conditions ([8]; *see* **Notes 21, 22**).
4. Basic quality control includes equal seeding (Fig. 1c), expected starting ratio (*see* **Note 23**), response to control compounds (*see* **Note 24**), and consistency in replicates (Fig. 1d).
5. The visual appearance of cells can help to further classify hit compounds [24].

4 Notes

1. We have very good experience with this retroviral vector (access to S2 facility needed). There are several alternative LC3B expressing vectors available on Addgene and the literature, not depending on S2 facilities. We recommend the use of ratiometric fluorescent reporters (GFP-mCherry/RFP double tag, mKeima, etc.).

2. The use of an adherent cell line stably expressing the reporter is highly preferable. Use of non-adherent cell lines is possible; however, we experienced higher variations using these cell lines and therefore considerably decreased sensitivity. Our laboratory uses among others the RPE-1 cell line (ATCC nr. CRL-4000), which is an immortalized, non-disease, epithelial cell line with low basic autophagy flux. Cancer cell lines tend to have a high basic autophagy flux level.
3. DMSO is a universal solvent for many compounds. Concentrations of <0.1% DMSO (v/v) are non-toxic (no detectable induction of autophagy or growth delay) for most of the commonly used immortalized cell lines. For more sensitive cell lines, this may differ and needs to be tested before the beginning of the screen.

   Torin 1 is a potent inhibitor of the mTOR complex and therefore an inducer of autophagy. We found 250 nM Torin 1 sufficient for the max. Induction of autophagy in most cell lines. This concentration will cause a strong reduction in cell proliferation without induction of cell death in most cell lines. In case of doubt, respective pretests should be performed to find the optimal/maximal concentration.

   Bafilomycin A1 (Baf A1) is an inhibitor of the autophagy pathway. Cell lines dependent on autophagy may undergo increased cell death in the presence of Baf A1 (cell line specific).
4. Aliquot the prepared compound solutions in, e.g., PCR tubes (Torin 1, bafilomycin A1) and use a fresh aliquot for each preparation.
5. Control compounds need to be adapted to the solvent of compounds used in the screen (DMSO, ethanol, etc.). Stock concentrations of compounds should be ideally >1000x to avoid an impact of the solvent on autophagy flux (*see* **Note 3**).
6. We did not observe a significant difference between the usage of more expensive screening plates from other companies and cost-saving.
7. Autophagy is a stress response pathway. Make sure your cell lines are mycoplasma-free. Do not use freshly thawed (<8 days in culture and split for <2x) or over−/undergrown cells. Confluency of cells used for the screen/seeding should be consistent to ensure the reproducibility of results (~80% confluent).
8. We use a standard final volume of 100 μl for screens lasting 72 h and seed cells in all 384 wells. Less volume can be used to save material/compounds. We recommend seeding cells in a volume of >25 μl and to perform the screen in a final volume of >30 μl. Using low volumes abet the occurrence of edge effects (evaporation). This can be minimized by higher volumes

(100 μl) of media or PBS in the two unused outer lines of wells (see plate map).

9. Prepare 35–40 ml of cell suspension per plate. Cells sediment-fast: Gently invert the tube of cell suspension 10–15 times right before starting the seeding process to ensure equal seeding of cells. Flush the tubes of the Multidrop™ properly with cell suspension before starting with the plate (death volume ~7 ml). Insufficient trypsinization can lead to cell clumps and also be the cause of unequal seeding.
10. Touch the plate only at the border area to avoid fingerprints on the screening area (lid and bottom) as this may interfere with automated focusing.
11. The temperature of the cell suspension drops during the seeding process. Transfer of the plate back to the incubator (37 °C) induces mild convection of the media in individual wells. Therefore, cells need to sediment before being transferred to the incubator to avoid uneven distribution within individual wells (accumulation of cells at the center of the well).
12. The optimal cell seeding concentration and growth time are cell line-dependent and slower-growing cells need a higher initial concentration and/or longer growth time. We recommend having ~20% confluency at the time of treatment. For example, U2OS cells need 30.000 cells/ml, while HEK293T cells need 40.000 cells/ml, to reach 20% confluency after 24 h. Cells at low confluency are usually more sensitive to compounds and high confluency may mask hits.
13. Using 384-well plates and the Echo® 550, compounds can be directly added to the screening plate on day 2. Remember to adapt the volume of media during cell seeding and/or amount of compound accordingly.
14. Thaw compound source plates for sufficient time, shake (resolve), and centrifuge before removing the sealing to avoid cross-contamination and pipetting errors. Pay attention to the potential crystallization of compounds over time (resolve using e.g., ultrasound).
15. Controls are needed for the interpretation of the compound effect and for quality control. For the latter, it is important to distribute controls over the plate to be able to identify potential edge effects (evaporation; temperature) and plate production errors.
16. Standard scan program: magnitude of 10x; green, red, and phase channel; scan every 2 h over a period of 72 h. In case of your conditions, do not fill the complete plate, and scan only treated wells to minimize screening time (full plate/60 conditions: ~10 min). With this setting, six plates (better only five to

avoid overheating) can be measured in parallel in the Incucyte®. The time of treatment of individual plates needs to be coordinated, respectively, so that the time between treatment and the first scan of a plate is consistent.

17. Bubbles may interfere with the automated focus process. To remove bubbles, blow air on them with a pissette wash bottle containing a bit of ethanol. Raise the dispenser in such a way that it is not in contact with the ethanol. First, make a test blowing on your hand so that no liquid ethanol comes out.
18. The waiting time is needed to allow potential condensation at the lid to dissolve. Condensation at the lid due to temperature change may disturb the automated focus process. When screening more than one plate, the plate can be incubated within the incubator but outside the Incucyte® while scans are in progress.
19. The setup for measuring confluency is cell line-dependent. In exceptional cases, the phase contrast of very flat cells may be insufficient for this kind of analysis.
20. See supplementary Excel file including a standard analysis pipeline.
21. The ratiometric character of the analysis allows comparison between conditions independent of the cell number. Normalization to the confluency is not necessary. Therefore, a plate reader is in principle sufficient to acquire data for autophagy flux analysis. At the same time, measurements are performed outside an incubator, which can stress cells and thereby impact autophagy flux (moving cells, drop in temperature and CO_2 levels). Therefore, fast handling is important. Be careful, if you plan to normalize the green/red fluorescence data to the first time point. Fast-acting compounds may shift the ratio at time point 0 and normalization to the first time point may mask this. It is better to look at the ratios without normalization.
22. Dependent on the used reporter (see introduction), analyzing the red/green ratio is to be preferred. This is, e.g., the case for several ER-phagy and mitophagy reporters.
23. Each created reporter cell line will have its specific ratio under basal, non-stressed conditions, which will stay constant over time. It depends on the basal autophagy flux of the specific cell line and on the expression level of the reporter. High expression levels of a reporter may already induce stress response pathways; therefore, low-to-medium expression that still gives a sufficient signal for analysis is preferred during the generation of stable reporter cell lines. Change/difference in the starting ratio of a reporter line indicates stress (mycoplasma, cells have been overgrown, cells are too old, etc.). This impacts the

outcome of a screen, making it non-comparable/incorrect, and therefore, cells/data should be discarded.

24. Stable reporter cell lines will have a specific response pattern to control compounds, which should be established prior to the screen. Aberrations (e.g., less response to Torin 1) indicate edge effects (only part of the control compounds are affected), stressed cells (*see* **Note 23**), or old control compounds (*see* **Note 4**).

Acknowledgments

This work was supported by the German Research Foundation DFG (SFB1177/2 and WO210/20-2), the Dr. Rolf M. Schwiete Stiftung (13/2017), the Hessian Ministry of Science and Art (HMWK) initiative ENABLE, and the EU/EFPIA/OICR/McGill/KTH/Diamond Innovative Medicines Initiative 2 Joint Undertaking (EUbOPEN Grant No. 875510).

References

1. Levine B, Kroemer G (2019) Biological functions of autophagy genes: a disease perspective. Cell 176:11–42
2. Dikic I, Elazar Z (2018) Mechanism and medical implications of mammalian autophagy. Nat Rev Mol Cell Biol 19:349–364
3. Klionsky DJ et al (2021) Autophagy in major human diseases. EMBO J 40:e108863
4. Deretic V, Kroemer G (2022) Autophagy in metabolism and quality control: opposing, complementary or interlinked functions? Autophagy 18:283–292
5. Mizushima N, Murphy LO (2020) Autophagy assays for biological discovery and therapeutic development. Trends Biochem Sci 45:1080–1093
6. Ueno T, Komatsu M (2020) Monitoring autophagy flux and activity: principles and applications. BioEssays 42:e2000122
7. Mizushima N, Yamamoto A, Matsui M, Yoshimori T, Ohsumi Y (2004) In vivo analysis of autophagy in response to nutrient starvation using transgenic mice expressing a fluorescent autophagosome marker. Mol Biol Cell 15: 1101–1111
8. Kaizuka T et al (2016) An autophagic flux probe that releases an internal control. Mol Cell 64:835–849
9. Kimura S, Noda T, Yoshimori T (2007) Dissection of the autophagosome maturation process by a novel reporter protein, tandem fluorescent-tagged LC3. Autophagy 3:452–460
10. Pankiv S et al (2007) p62/SQSTM1 binds directly to Atg8/LC3 to facilitate degradation of ubiquitinated protein aggregates by autophagy. J Biol Chem 282:24131–24145
11. Khaminets A et al (2015) Regulation of endoplasmic reticulum turnover by selective autophagy. Nature 522:354–358
12. Liang JR, Lingeman E, Ahmed S, Corn JE (2018) Atlastins remodel the endoplasmic reticulum for selective autophagy. J Cell Biol 217:3354–3367
13. Liang JR et al (2020) A genome-wide ER-phagy screen highlights key roles of mitochondrial metabolism and ER-resident UFMylation. Cell 180:1160–1177.e20
14. An H et al (2019) TEX264 is an endoplasmic reticulum-resident ATG8-interacting protein critical for ER remodeling during nutrient stress. Mol Cell 74:891–908.e10
15. Chino H, Hatta T, Natsume T, Mizushima N (2019) Intrinsically disordered protein TEX264 mediates ER-phagy. Mol Cell 74: 909–921.e6
16. Reggio A et al (2021) Role of FAM134 paralogues in endoplasmic reticulum remodeling, ER-phagy, and collagen quality control. EMBO Rep 22:1–20

17. Chen Q et al (2019) ATL3 is a tubular ER-Phagy receptor for GABARAP-mediated selective autophagy. Curr Biol 29:846–855.e6
18. Katayama H, Kogure T, Mizushima N, Yoshimori T, Miyawaki A (2011) A sensitive and quantitative technique for detecting autophagic events based on lysosomal delivery. Chem Biol 18:1042–1052
19. Katayama H et al (2020) Visualizing and modulating mitophagy for therapeutic studies of neurodegeneration. Cell 181:1176–1187.e16
20. Allen GFG, Toth R, James J, Ganley IG (2013) Loss of iron triggers PINK1/parkin-independent mitophagy. EMBO Rep 14: 1127–1135
21. McWilliams TG et al (2016) Mito-QC illuminates mitophagy and mitochondrial architecture in vivo. J Cell Biol 214:333–345
22. Eapen VV, Swarup S, Hoyer MJ, Paulo JA, Harper JW (2021) Quantitative proteomics reveals the selectivity of ubiquitin-binding autophagy receptors in the turnover of damaged lysosomes by lysophagy. elife 10
23. Diehl V et al (2021) Minimized combinatorial CRISPR screens identify genetic interactions in autophagy. Nucleic Acids Res 49:5684–5704
24. Wells CI et al (2021) The Kinase Chemogenomic Set (KCGS): an open science resource for kinase vulnerability identification. Int J Mol Sci 22

Chapter 17

Phenotypic Chemical Screening in CD4$^+$ T Cells to Identify Epigenetic Inhibitors

Adam P. Cribbs and Udo Oppermann

Abstract

Chemical biology provides an attractive approach to identify genes involved in a particular biological process. This screening approach has its advantages because the assays are usually non-destructive, and analysis can be performed even if the mechanism of action is unknown. During an immune reaction, cells upregulate the expression and secretion of small proteins called cytokines that have specific effects on the interactions and communication between cells. Here, we describe the principles and steps involved in the execution of chemical screening for identifying epigenetic inhibitors that affect cytokine production in differentiated Th1, Th2, and Th17 cells. Our approach provides a rationale for identifying epigenetic chemical compounds that are capable of controlling CD4$^+$ T-cell cytokine function that may be beneficial for treating inflammatory diseases.

Key words Epigenetic inhibitor, Immune cells, Cytokines, T helper cell

1 Introduction

We here outline a protocol for chemical screening that involves the selection of chemical probes based on their ability to induce "epigenetic" changes (i.e., use of selective molecules targeting so-called readers, writers, and erasers of a histone or chromatin code). The protocol is applicable to other focused screening efforts with chemical libraries to identify changes that can occur following the treatment of cells or samples [1–4]. Screens are typically non-destructive, allowing for an understanding of chemical-induced phenotypic changes, followed by destructive approaches to investigate mechanism, such as genetic analyses being performed within the same sample. Phenotypic screens have their advantages because the analysis can be performed even if the mechanism of action is unknown. However, selecting the appropriate phenotypic readout can be challenging as it is necessary to choose the characteristic that best reflects the function of the system you are

Daniel Merk and Apirat Chaikuad (eds.), *Chemogenomics: Methods and Protocols*, Methods in Molecular Biology, vol. 2706, https://doi.org/10.1007/978-1-0716-3397-7_17,

studying. In the current chapter, we describe how phenotypic-based chemical screening can be used to identify compounds that regulate cellular function. Specifically, we provide the example of how epigenetic compound screening can be employed to identify targets that impact Th1, Th2, and Th17 cell function [5].

Cytokines are small secreted proteins released by cells that have specific effect on the interactions and communications between cells. In inflammatory responses, cytokines are released, which act to stimulate, recruit, and proliferate immune cells. The inflammatory response is coordinated by a time-dependent secretion of different cytokines and recruitment of different types of immune cells. Typically, myeloid cells drive early inflammation [6], while lymphoid cells are typically seen during the latter stages of inflammation. Autoimmune diseases are characterized by sustained levels of chronic inflammation that fail to resolve, which is explained in part by high levels of $CD4^+$ T-cell infiltration and cytokine expression [7]. Upon activation, naïve $CD4^+$ T cells differentiate into T helper subsets, including Th1, Th2, and Th17, which are each characterized by different cytokine profiles [8]. Despite many successful anti-cytokine treatments for autoimmune diseases, there are groups of patients that do not respond to therapy. Moreover, certain autoimmune diseases such as systemic lupus erythematosus (SLE) are still difficult to treat using biologic drugs [9]. This provides a rationale for identifying chemical compounds that are capable of controlling $CD4^+$ T-cell cytokine function that may be beneficial for treating inflammatory diseases, such as SLE.

Here, we present the principles and steps involved in execution of phenotypic chemical screening for epigenetic compounds that affect cytokine expression in Th1, Th2, and Th17 cells. The compound library comprises inhibitors of "writers" such as histone methyl- or acyl-transferases, "readers" such as acetyl or methyl-binding chromatin domains, and "erasers" such as histone demethylases or deacetylases. We refer the readers to selected examples in other areas such as cancer cell proliferation where phenotypic screens helped to identify novel targets in rare diseases such as chordoma [10] or in stem cell biology by identifying molecular barriers in somatic cell reprogramming to induced pluripotent stem cells [11, 12].

2 Materials

2.1 Isolating Mononuclear Cells

1. Blood obtained from platelet apheresis residues or from venous draws from healthy volunteers.
2. Ficoll-Paque.
3. 50 mL Falcon tubes.

2.2 Isolating Naïve CD4⁺ T Lymphocytes

1. Naïve CD4⁺ T-cell Isolation Kit II (Miltenyi).
2. Magnetic separation stand and magnet (Miltenyi).
3. MACS buffer: PBS supplemented with 0.5% FCS and 2 mM ethylenediaminetetraacetic (EDTA).
4. 15 ml Falcon tubes.

2.3 Culture and Differentiation Conditions

1. Culture media: X-VIVO containing 10% knockout serum replacement (Thermo Fisher).
2. Anti-CD3/CD28 activation beads (Invitrogen).
3. Cytokines: recombinant human IL-12 (R&D), recombinant human IL-4 (PeproTech), recombinant human IL-1β (PeproTech), recombinant human TGF-β (PeproTech), and recombinant human IL-23 (R&D).
4. Antibodies: anti-IL-4 neutralizing antibody (R&D) and anti-IFN-γ neutralizing antibody.
5. 96-well round-bottom cell culture quality polypropylene plate.

2.4 Chemical Library

1. 96-well flat bottom culture quality polypropylene plates.
2. Dimethyl sulfoxide (DMSO).
3. Aluminum plate sealing tape.
4. Chemical library: This library consists of 226 small-molecule epigenetic inhibitors (Fig. 1). The library is arrayed into a 96-well plate format at concentrations between 0.05 and 1000 mM pre-dissolved in 100% DMSO and is stored at −20 °C [5]. The plate is used as a master plate, and preparation of working solutions is made by diluting the library 1 in 1000 with 100% DMSO.

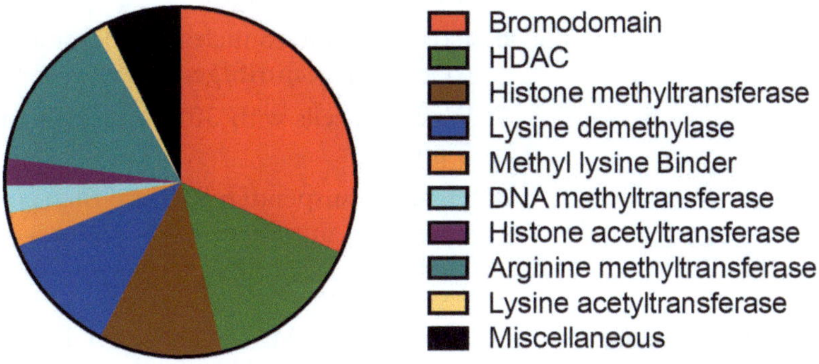

Fig. 1 The proportions of epigenetic inhibitor compounds showing the distribution by compound type

2.5 ELISAs

1. 96-well ELISA plates.
2. Human IFN-γ ELISA Ready-SET-Go®.
3. Human IL-17 ELISA Ready-SET-Go®.
4. Human IL-4 ELISA Ready-SET-Go®.
5. FLUOstar Omega Plate Reader capable of measuring absorbance (BMG Labtech).

2.6 Cell Viability

1. alamarBlue™ Cell Viability Dye (Thermo Fisher).
2. FLUOstar Omega Plate Reader capable of measuring fluorescence (BMG Labtech).

3 Methods

All incubations are at 37 °C in an incubator containing 5% CO_2, unless otherwise stated.

3.1 Isolating Mononuclear Cells

Blood should be collected in compliance with local ethics guidelines. Waste blood products should be disposed of in accordance with local rules.

1. Invert the Ficoll-Paque media bottle several times to ensure a good mix (*see* **Note 1**).
2. Add either 3 mL (if working with venous draw blood) or 20 mL (if working with blood obtained from platelet apheresis residues) into a 15-mL or 50-mL centrifuge tube.
3. Dilute blood to an appropriate volume with PBS so that it can be carefully layered over the Ficoll-Paque.
4. Centrifuge the tube at 400 g for 25 min at 18–20 °C (**Note 2**).
5. Draw the upper layer containing the plasma and platelets and discard appropriately.
6. Isolate the mononuclear cell layer at the interface and then add to a 50-mL centrifuge tube.
7. Wash the cells with 30 mL of PBS and centrifuge at 800 g for 5 mins.
8. Finally, resuspend in PBS and then count the cell concentration using an automatic cell counter or hemocytometer (see **Note 3**).

3.2 Isolating Naïve CD4$^+$ T Lymphocytes

1. Add 40 μL of MACS buffer per 1×10^7 cells, then add 10 μL of naïve CD4$^+$ T-cell biotin antibody cocktail II per 1×10^7 cells, and incubate for 5 mins at 4 °C (*see* **Note 4**).
2. Add 30 μL of MACS buffer per 1×10^7 cells, add 20 μL of naïve CD4$^+$ T-cell MicroBead Cocktail II per 1×10^7 cells, and incubate for 10 mins at 4 °C.

3. Place an LS column into a MidiMACS separator and wash with 3 mL of MACS buffer.
4. Add the mononuclear cells containing antibody to the column and wash with 3 mL of MACS buffer.
5. Collect the flow-through containing unlabeled naïve $CD4^+$ T cells (*see* **Note 5**).
6. Resuspend cells in X-VIVO 15 medium supplemented with 10% serum replacement at a concentration of 1×10^6 cells/mL.

3.3 Culture and Differentiation Conditions

1. To differentiate enough cells for large screening assays (greater than 500 compounds), we recommend using a minimum of 20×10^6 naïve cells to ensure enough differentiated cells at day 6.
2. Th2 differentiation: Culture naïve $CD4^+$ T cells in the presence of recombinant human IL-4 (4 ng/mL) and anti-IFN-γ neutralizing antibody (10 mg/mL).
3. Th1 differentiation: Culture naïve $CD4^+$ T cells in the presence of recombinant human IL-12 (5 ng/mL) and anti-IL-4 neutralizing antibody (10 mg/mL).
4. Th17 differentiation: Culture naïve $CD4^+$ T cells in the presence of recombinant human IL-1β (20 ng/mL), recombinant human TGFβ (3 ng/mL), and recombinant human IL-23 (10 ng/mL).
5. Culture cells for a further 6 days (*see* **Note 6**).

3.4 Chemical Library

1. On day 6 of the differentiation process, wash cells in fresh X-VIVO 15 medium and then count cells.
2. Plate 50,000 cells per well into 69-well plates in 180 μL of X-VIVO 15 media.
3. Thaw the master plate containing compounds at high concentrations and dilute 1 in 100 with fresh X-VIVO 15 media.
4. Add 20 μL of this intermediate dilution to the culture wells containing differentiated cells (*see* **Note 7**).
5. Culture the cells for 24 h.

3.5 ELISAs

1. Following 24 h of compound treatment, harvest the supernatant and add it to a separate 96-well plate (*see* **Note 8**).
2. Add 90 μL of fresh medium to the 96-well plate containing cells and perform cell viability assay (*see* Subheading 3.6).
3. Perform ELISA on the supernatants to measure the expression of IFN-γ from Th1 cells, IL-4 from Th2 cells, and IL-17 from Th17 cells.
4. Calculate background from media-only wells and subtract from all other wells. Compare cytokine changes in response to compound treatment to DMSO-only controls.

3.6 Cell Viability

1. Add 10 μL of alamarBlue to the 96-well plate containing cells.
2. Incubate the cells for 1–4 h (see **Note 9**).
3. Read the fluorescence of the plate using a 96-well plate reader.
4. Normalize cytokine concentrations against cell viability.

4 Notes

1. Warm the Ficoll-Paque density gradient media to 18–20 °C prior to use.
2. Ensure the break of the centrifuge is turned off.
3. Washing will remove most platelets; however, if there is a high platelet concentration, then centrifugation should be performed at 300 g for 15 min and then cells re-counted. If there is a high dead cell count, then the cells should be resuspended in 50 mL of PBS and then layered over two 25 mL of Ficoll-Paque.
4. It is important to follow the manufacturer's recommended incubation times precisely to obtain a high naïve $CD4^+$ T-cell population.
5. The purity of naïve $CD4^+$ T cells should be 90–98%, which can be verified using flow cytometry after staining for CD45RA and CD45RO.
6. In other iterations of the protocol, fresh cytokines can be added after day 3.
7. Compare compound effects with cells cultured in 0.1% DMSO alone and use wells filled with media alone as control blanks for correcting background.
8. The supernatant can be stored long-term at −20 °C.
9. The wells that contain highly proliferative cells turn purple. It is advisable that the place is initially read every 30 mins to determine the appropriate incubation time.

Acknowledgements

This work was supported through Innovate UK (UO, APC), the National Institute for Health Research Oxford Biomedical Research Centre (UO), Cancer Research UK (CRUK, UO), the Bone Cancer Research Trust (APC and UO), the Leducq Epigenetics of Atherosclerosis Network (LEAN) program grant from the Leducq Foundation (UO), the Chan Zuckerberg Initiative (APC), and the Myeloma Single Cell Consortium (UO). APC is a recipient of an MRC Career Development Fellowship (MR/V010182/1).

References

1. Khersonsky SM, Chang YT (2004) Forward chemical genetics: library scaffold design. Comb Chem High Throughput Screen 7(7): 645–652
2. Zheng W, Thorne N, McKew JC (2013) Phenotypic screens as a renewed approach for drug discovery. Drug Discov Today 18(21–22): 1067–1073
3. Choi H et al (2014) Forward chemical genetic screening. Methods Mol Biol 1062:393–404
4. Chen GQ et al (2019) Phenotype and target-based chemical biology investigations in cancers. Natl Sci Rev 6(6):1111–1127
5. Cribbs AP et al (2020) Histone H3K27me3 demethylases regulate human Th17 cell development and effector functions by impacting on metabolism. Proc Natl Acad Sci U S A 117(11):6056–6066
6. Shi C, Pamer EG (2011) Monocyte recruitment during infection and inflammation. Nat Rev Immunol 11(11):762–774
7. Skapenko A et al (2005) The role of the T cell in autoimmune inflammation. Arthritis Res Ther 7(Suppl 2):S4–S14
8. Sandquist I, Kolls J (2018) Update on regulation and effector functions of Th17 cells. F1000Res 7:205
9. Samotij D, Reich A (2019) Biologics in the treatment of lupus erythematosus: a critical literature review. Biomed Res Int 2019: 8142368
10. Cottone L et al (2020) Inhibition of histone H3K27 demethylases inactivates Brachyury (TBXT) and promotes Chordoma cell death. Cancer Res 80(20):4540–4551
11. Ebrahimi A et al (2019) Bromodomain inhibition of the coactivators CBP/EP300 facilitate cellular reprogramming. Nat Chem Biol 15(5): 519–528
12. Ugurlu-Cimen D et al (2021) AF10 (MLLT10) prevents somatic cell reprogramming through regulation of DOT1L-mediated H3K79 methylation. Epigenetics Chromatin 14(1):32

Index

Daniel Merk and Apirat Chaikuad (eds.), *Chemogenomics: Methods and Protocols*, Methods in Molecular Biology, vol. 2706, https://doi.org/10.1007/978-1-0716-3397-7,

M

N

P

Q

S

T

U

V